DATE DUE

ENGINEERING TOMORROW

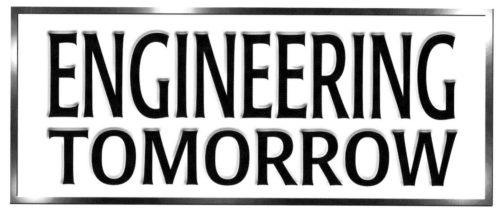

ENGINEERING TOMORROW

Today's Technology Experts
Envision the Next Century

Janie Fouke, Editor

Trudy E. Bell and Dave Dooling, Writers

IEEE
PRESS

This book and other books may be purchased at a discount from the publisher when ordered in bulk quantities. Contact:

IEEE Press Marketing
Special Sales
P.O. Box 1331, 445 Hoes Lane
Piscataway, NJ 08855-1331
Fax: +1 732 981 9334

For more information about IEEE Press products, visit the IEEE Press Home Page:
http://www.ieee.org/press

Printed in Canada

10   9   8   7   6   5   4   3   2   1

ISBN 0-7803-5360-9 (Platinum Edition)   IEEE Order Number: PC5801
ISBN 0-7803-5361-7 (Member cloth edition) IEEE Order Number: PC5802
ISBN 0-7803-5362-5 (Trade cloth edition)   IEEE Order Number: PC5803

Publishers Cataloging-in-Publication Data

Bell, Trudy E.
    Engineering tomorrow : today's technology
experts envision the next century / Janie Fouke,
editor ; Trudy E. Bell and Dave Dooling, writers.
-- 1st ed.
    p. cm.
    Includes index.
    ISBN: 0-7803-5360-9 (limited platinum ed.)
    ISBN: 0-7803-5361-7 (IEEE member cloth ed.)
    ISBN: 0-7803-5362-5 (trade cloth ed.)

    1. Technological forecasting. 2. Twenty-first
century--Forecasts. 3. Engineering--Forecasting.
I. Dooling, David. II. Fouke, Janie. III. Title.

T174.B45 1999   303.48'3
                                 QBI99-1267

Cover Art

artist:
"Suntrapolis" by Mark Overs
Stainless Steel (36" x 30" x 30")
represented by:
Bilhenry Gallery, Milwaukee, WI
(414) 332-2509 • www.execpc.com/~bilhenry/bhgalry.html

The Jester, p. 124, courtesy of Digital Domain, Inc.
"Tightrope" © 1998 Digital Domain, Inc. All rights reserved.

**Editorial and production services provided by
Laing Communications Inc., Redmond, Washington.**

Book design and production: Sandra J. Harner
Cover design and spot illustrations: Kelly C. Rush
Editorial and production management: Laura B. Fisher

TABLE OF CONTENTS

Foreword • xi

Preface • xiii

Chapter 1: Threshold of the New Millennium • 1

Vinton G. Cerf: What are key policy and social issues facing the Internet? • 4

Wilson Greatbatch: What is the secret of happiness in a career? • 8

Jack S. Kilby: What are some of technology's unanticipated consequences? • 12

Arno A. Penzias: What constitutes a good scientific theory? • 18

Charles H. Townes: How can we be wiser about decisions of what to pursue in science and technology? • 24

PART I—TECHNOLOGIES FOR SOCIETY'S INFRASTRUCTURE

Chapter 2: Structures and Devices • 31

John G. Kassakian: How soon can we free ourselves from fossil fuels? • 34

Roger D. Pollard: If you could 'uninvent' a technology, which would it be? • 40

Donald R. Scifres: How will information technology transform global culture? • 44

Ralph C. Merkle: Are we prepared for the nanotechnology revolution? • 50

Chapter 3: Systems and Management • 55

Robert A. Bell: When will society recognize that nuclear reactors are environmentally safer than fossil-fuel power plants? • 58

Samuel J. Keene: How can effective communication help engineers develop the best products? • 64

Wade H. Shaw, Jr.: Can engineers abdicate leadership forever? • 70

Rui de Figueredo: How can we accurately evaluate creativity and diversity? • 74

Chapter 4: Computers and Software • 79

Douglas C. Engelbart: Can we make society smarter? • 82

Alan Kay: How can we separate the Internet's wheat from its chaff? • 88

Gordon E. Moore: Are computers really the tide that will float all boats? • 94

Rao R. Tummala: How can we prevent ourselves from drowning in high-tech waste? • 98

PART II—HUMAN APPLICATION TECHNOLOGIES

Chapter 5: Communications • 103

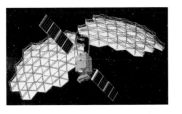

Tingye Li: Are we eating our seed corn? • 106

Rod C. Alferness: When is unlimited information effectively no information at all? • 110

Bennett Z. Kobb: The electromagnetic spectrum—public trust or pork barrel? • 114

A. Robert Calderbank: What is the role of industrial research laboratories in the twenty-first century? • 120

Chapter 6: Entertainment • 125

Wayne C. Luplow: What is the most environmentally sound way to dispose of consumer electronics products? • 128

Stephen B. Weinstein: How will the Internet affect social relationships? • 132

S. Joseph Campanella: What is the future of the U.S.'s universities and corporate research laboratories? • 136

Robert W. Lucky: Why can't we better predict which technologies will succeed? • 140

Chapter 7: Medicine and Biology • 145

Cato T. Laurencin: How can mentoring overcome racial discrimination? • 148

Thelma Estrin: What practical advice can encourage women engineers? • 154

George S. Moschytz: How can we ensure that technology is humane and not inane? • 160

Ray Kurzweil: When computer intelligence exceeds human intelligence, what will it mean to be human? • 166

Chapter 8: Transportation • 171

Victor Wouk: How much will we pay for freedom of movement? • 174

Linda Sue Boehmer: What is the potential of computer intelligence in mass transit? • 178

William F. Powers: Will cars ever have jet fighter controls? • 184

George L. Donahue: Can we overcome our fear of flying? • 188

Chapter 9: Exploration • 193

Freeman J. Dyson: How can we further explore the 'microverse'? • 196

Burt Rutan: Why are humans driven to explore? • 200

Robert Zubrin: Do we really need an armada to explore Mars? • 204

Joseph R. Vadus: Will humans live in cities floating on the oceans? • 208

PART III—ENGINEERING OUR PRIORITIES

Chapter 10: The Environment • 215

Stewart Brand: Why should engineers take the long view? • 218

Ghassem Asrar: How can we best invest in the next generation of scientists and engineers? • 222

Sylvia A. Earle: What does it take for people to realize that technology-induced climate change is jeopardizing our very lives? • 228

M. Granger Morgan: Why is it urgent now to investigate low-carbon sources of energy? • 234

Chapter 11: War and Peace • 241

Norman R. Augustine: How can we watch out for a weapon that hasn't been invented? • 244

Richard L. Garwin: Can we mount an effective defense without having to shoot? • 248

David A. Kay: How much privacy will we trade for safety? • 254

Myron Kayton: How do we reduce the body count? • 260

Chapter 12: Preparing Engineers for Tomorrow • 265

Edward Alton Parrish: How can students experience the impact of engineering on society? • 268

Donald Christiansen: Engineering ethics—who cares? • 272

Donna Shirley: Why is diversity essential to sustaining creativity? • 278

John B. Slaughter: How can people learn to get along better? • 284

Eleanor Baum: How can more young people be attracted to engineering? • 288

Appendix: The Fifty Technology Experts • 293

About the Authors • 297

About the Editor • 298

Acknowledgments • 299

Index • 302

FOREWORD

Engineering with a Conscience

LET'S THINK VERY, VERY CAREFULLY about how we engineer tomorrow.

That is the central message from all 50 technology experts featured in *Engineering Tomorrow: Today's Technology Experts Envision the Next Century*.

"How can we ensure that technology is humane and not inane?" probes George S. Moschytz (Swiss Federal Institute of Technology, Zurich, Switzerland). The touchstone for judgment, in his view, is whether a product or project aids humanity. That may mean making things that may not in themselves be wildly profitable—and not making other things, profitable or not, because they yield no positive good to individual humans or to society at large. "Just because something can be done does not necessarily mean it should be done," he cautions.

"Engineers can invent almost anything society wants," agrees Roger D. Pollard (University of Leeds, England). "The question is, is society going to make good use of it?"

Pithiest of all is the summation of Wilson Greatbatch (inventor of the fully implantable cardiac pacemaker): "You should do your work because it is a *good* thing to do."

That engineers have a positive social responsibility to make ethical judgments about goodness—and that, indeed, engineering is a helping profession as surely as medicine or education—is a repeating theme among these Nobel Prize laureates, engineering society fellows, and other luminaries. While their fields of interest range from subatomic particles (Freeman J. Dyson, Institute of Advanced Study, Princeton, New Jersey) to exploration of the Red Planet (Robert Zubrin, Mars Society, Indian Hills, Colorado), more than half of their matters of conscience fall into one of four general categories. Numerically most prevalent are concerns about the environment in the twenty-first century—specifically about clean sources of energy and the mitigation of human-induced global warming. Right behind are concerns about the future of basic research, the societal and cultural implications of the global Internet, and the provision of adequate education for all children and young adults—specifically for the disenfranchised (women, minorities, handicapped, and economically disadvantaged).

Some of these issues they see as being urgent matters of life and death. "What does it take for people to realize that technology-induced climate change is jeopardizing our very lives?" asks Sylvia Earle (National Geographic Society, Washington, D.C.). "The extent to which we alter the natural system has profound consequences to our immediate future—to the lifetime most of us can look forward to having, not just to the future of our children and beyond."

Many issues are complex and long-term, and their eventual solution will require that not only individual engineers but also corporations and other institutions act in good conscience. Indeed, engineering with a conscience is just plain good, long-range business sense, points out Stewart Brand (Global Business Network, Emeryville, California; founder of *The Whole Earth Catalog* and the *CoEvolution Quarterly*). As counter-intuitive as it may sound, Brand points out, "if you look farther ahead into the future than just the next quarter or the next year, things that look like altruism actually become self-interest."

One universal solution—literally all the way from product design to world peace—is just for humans to talk openly with one another. "The biggest problem in product development is communication," declares Samuel J. Keene (Performance Technology, Boulder, Colorado). "It's also the seed of the solution." Keeping technical capabilities as visible as possible is the best defense against warfare, argues Richard L. Garwin (IBM Fellow Emeritus, Thomas J. Watson Research Center, Yorktown Heights, New York); "You want to dissuade people from behaving in an inimical fashion." When people have the opportunity "to communicate across would-be barriers of race, culture, and language, they establish understanding," observes John B. Slaughter (Occidental College, Los Angeles, California). "The barriers disappear, and they discover there is so much more they can do in collaboration."

And such communication must start as early as possible, states Edward Alton Parrish (Worcester Polytechnic Institute, Worcester, Massachusetts)—preferably in college undergraduate years or even earlier. "We need to make students sensitive about the impact of engineering on society—even if they don't have to meet environmental or other codes for their type of project."

"It's important that engineers not just build stuff, but that they build the right stuff," adds Brand. "In the twenty-first century, it's imperative that they see how the content fits into the broader environmental and cultural context. So in the training of engineers—more history, please, and more humanities."

According to Sylvia Earle, two simple technologies can remind us who is ultimately responsible for engineering our future and what it takes to get started. "One is a mirror. Everyone ought to pick one up and look hard in it. Don't wait for the next guy to take action. The other is a two-by-four, that stick that smart farmers in the U.S. South use to whack mules between the ears to get their attention. We need to do that to ourselves, and say, 'Get busy.'"

—Janie Fouke, Ph.D.
Dean, College of Engineering
Michigan State University
East Lansing, Michigan

PREFACE

OVER THE LAST MILLENNIUM, THE TWENTIETH CENTURY has ushered in more material advances than the eleventh through the nineteenth centuries combined. Most of those advances—the development of automobiles and aircraft, of computers and global communications, of minimally invasive surgery and medical imaging, and the setting of human and robotic feet on other planets—have been due to science and engineering.

In the next millennium, many significant challenges lie ahead. Some are as old as humanity itself—such as the provision of adequate housing, the production and distribution of food, and the availability of education and medical care. Others are new challenges resulting from developments unique to the twentieth century—such as the conservation of global resources, the control of global pollution, the maintenance of privacy in communications and computing, and the control of diseases unknown before the twentieth century.

Some of the challenges will require a purely technical approach. Others will be primarily societal, requiring a balancing of material priorities with sociological values—with the outcome expressed by funding availability and regulation.

All involve engineering techniques and expertise.

And never before have humans been so technically well-equipped.

Thus, on this, the dawn of the twenty-first century and the third millennium A.D., the Institute of Electrical and Electronics Engineers (IEEE) deems it fitting to pause and reflect on the roles of scientists and technologists in engineering tomorrow. As the world's largest engineering society with more than 340,000 members worldwide, the IEEE has roots in the past two centuries: the IEEE was formed in 1963 by a merger of the American Institute of Electrical Engineers (the electrical side, founded in 1884) and the Institute of Radio Engineers (the electronics side, founded in 1912).

It is, of course, electrical and electronic advances that have given the twentieth century so many of the technical wonders we take for granted. The advent of major grids for the distribution of electric power, for example, made urban subway systems and skyscrapers possible by encouraging the advent of elevators, water pumps, and huge-volume heating and air-conditioning systems. Similarly, the transistor—first as a discrete component and later as part of the integrated circuit—underlies every advance from mainframe and personal computers to cell phones and spacecraft. Entertainment, exploration, finance, food preparation and storage, medicine, military capabilities, transportation, the workplace, and even the education of children have all been transformed by advances in electrical and electronics engineering. In fact, some have argued that the inescapable pervasiveness of information technology—ranging from radio to fax machines to the Internet—has played a pivotal role in the dissolution of authoritarian governments, which can no longer keep their citizens in the dark.

Engineers have long been aware of their transformational influence on society's architecture. Although most young people are attracted to engineering because they love making technical concepts practical (and some because they want to try to get rich as entrepreneurs), many also want to help improve the quality of people's lives—and they see engineering as an effective way. The IEEE has had its Code of Ethics—an engineers' Hippocratic Oath for social responsibility—since 1979. And among its 40 constituent societies and technical councils, the IEEE has not only technical societies such as Computer, Communications, and Power Engineering (its three largest) but also "social conscience" societies such as Education, Engineering Management, and Social Implications of Technology.

At the millennial moment, this book is an informal "status report" on electrical and electronics engineering and its potential to help human society.

How the Book is Organized

Part I, "Technologies for Society's Infrastructure," highlights several significant devices, techniques, software developments, and systems concerns that will underlie many twenty-first-century advances. Part II, "Human Application Technologies," explores technologies that would directly affect individual humans (medicine and biology, entertainment) and human society (communication, transportation, and exploration). The book's final part, "Engineering Our Priorities," articulates principal issues in some of the largest technology-related challenges facing society (the environment, war and peace, and education).

One book of finite length cannot possibly address all seminal technologies or societal challenges. *Engineering Tomorrow* seeks primarily to identify key questions and opportunities, rather than to offer final answers. It highlights but a sampling of sociotechnical issues facing the twenty-first century—principal areas of concern and opportunity as seen from the vantage point of fifty of the twentieth century's top experts.

This book also seeks to open thoughtful conversations in government chambers, university lecture halls, and private living rooms. Engineers or not, we are all members of society. And in this listen-to-the-customer age, we vote our individual preferences for the future each time we use a credit card just as surely as when we cast a ballot.

Let the dialogue begin!

—Trudy E. Bell and Dave Dooling

CHAPTER 1

THRESHOLD OF THE NEW MILLENNIUM

THE NEW MILLENNIUM.

Will it be the best of times—when genetic engineering and "miracle" drugs and neural implants eradicate cancer, HIV, blindness, quadriplegia, and world hunger? Or will it be the worst of times—when irresponsible combustion of fossil fuels and smart bombs and neuroweapons lead to irreversible global climate change and new scourges of war, devastation, disease, and famine? Will instant worldwide communications liberate people in oppressed nations by arming them with

Left: This image of the Nile Delta, just north of Cairo, Egypt, encompasses the history of mankind's efforts at "engineering tomorrow." The delta's annual gift of fertile soil gave rise to the earliest human civilization and, with it, efforts to alter the environment. Little has changed over the past 6,000 years across much of the Nile. This image was taken October 4, 1994, by synthetic imaging radar—a joint U.S., German, and Italian effort—carried aboard the Space Shuttle. It reveals continued efforts to harness the Nile, often with technologies a few millennia old. More recent technologies, such as spaceborne radar, offer the potential for improved stewardship of the Earth. In this image, the Nile River splits into the Rosetta Branch, the curving dark line in the center of the image, and the Damietta Branch, the curving dark line in the lower right of the image. The light blue area on the right half of the image is a portion of the Nile River Delta. The thinner, straighter lines and the small network of gold lines are irrigation canals. A transition zone of irrigated fields is shown in blue and yellow between the irrigated delta and the surrounding desert. The desert is the dark blue area on the left side of the image lacking the pattern of irrigated fields. The 75-by-60-kilometer (46-by-37-mile) image is centered at 30.2° N, 31.1° E. North is toward the upper right. *NASA & Jet Propulsion Laboratory*

knowledge, power, and the public eye to halt atrocities? Or will omnipresent instantaneous communications lead to intrusive invasions of individual privacy and subtle new types of bondage and fraud?

What the future ultimately brings depends in large part on two basic factors: available technology and societal concern.

Technology provides the "how." Societal concern guides the "whether," the "what," the "why," the "when," the "whom," the "where," and the "how much."

In this book, some of the world's scientific and technological intellects address both technological and societal issues for the twenty-first century.

Despite a stereotype that prevails in some literature and popular culture that scientists and engineers are (at worst) mad or evil or (at best) unaware or unfeeling about the import of the inventions they "unleash" upon the world, reality is far different. As you will see throughout this chapter and the rest of this book, many technology experts are not only highly sensitive to the world's greatest ills and challenges, but are driven by urgency and compassion to right things for humanity as best they can.

They don't have all the answers.

Sometimes they don't have any answers at all—yet.

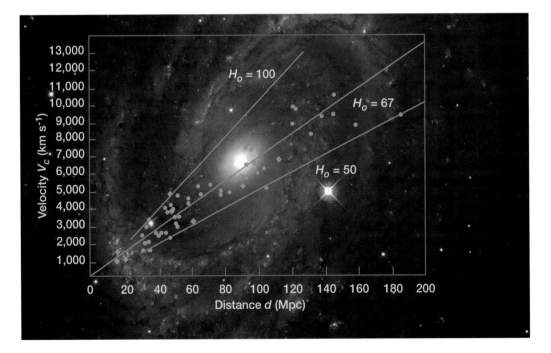

The cosmic microwave background showed that the universe is expanding almost uniformly in all directions. More recently, eight years of observations of "standard candle" stars—Cepheid variables—by the Hubble Space Telescope have shown that the rate of expansion increases with greater distance from the Earth. *NASA/Marshall Space Flight Center*

But they have identified many of the central, inescapable questions they feel will face human beings individually and collectively in the twenty-first century.

In this introductory chapter, you will meet five of the intellects whose contributions have helped sculpt the shape of technology and society in the last third of the twentieth century. Vinton G. Cerf co-invented the Internet. Wilson Greatbatch invented the first totally implantable self-powered biomedical device—the cardiac pacemaker. Jack S. Kilby invented the integrated circuit—the essential technology for the computer "chip." Arno A. Penzias co-discovered the 3 K cosmological background radiation—the first observational evidence in favor of a theory of the creation of the universe (the Big Bang). And Charles H. Townes co-invented the laser.

Each reflects a bit on the role of his development both technically and societally, and offers his individual personal perspective on its promises and challenges for the future.

You, dear reader, may find you either agree or disagree with the formulation of their questions or with the opinions they express.

Whichever is the case, *do* something about it. *Act* in the world according to your convictions and conscience.

That is the underlying message from all 50 luminaries interviewed for this book. Many have themselves dedicated a significant share of their lives to trying to better humanity's collective lot, through their technological developments, through broader social or political action, or both.

For they take to heart the warning of the British statesman Edmund Burke more than two centuries ago: "The only thing necessary for the triumph of evil is for good men to do nothing."

The Once and Future Internet

Talk about being in on the ground floor of a new technology. Vinton G. Cerf was a graduate student in computer science at the University of California at Los Angeles in 1968 when the Department of Defense's (DoD's) Advanced Research Projects Agency (ARPA) issued its RFQ (Request for Quotation) for a communications network that could link the computers of some 20 universities. The motivation was simple economics. ARPA, which was funding research in computer science, artificial intelligence, command and control and other information technologies that could potentially be "force multipliers" (allow a smaller force to triumph over a larger one in battle), could not afford to provide every university with a top-of-the-line computer. So a resource-sharing network, in which every university and the DoD would have access to every other's facilities, seemed to be a far more cost-effective approach.

VINTON G. CERF: What are key policy and social issues facing the Internet?

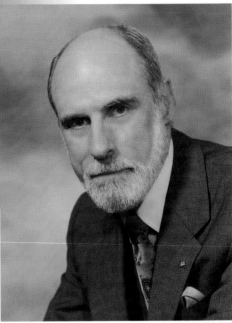

LOUIS FABIAN BACHRACH ©

"There are wide-ranging policy, social, and economic issues concerning the new medium of the Internet," stated Vinton G. Cerf, senior vice president of Internet architecture and technology at MCI WorldCom Corp. in Reston, Virginia. "I don't believe they'll be settled easily or quickly, but they must be addressed within the next few years."

Cerf is known as one of the two "fathers of the Internet"; the other is Robert E. Kahn (president of the Corporation for National Research Initiatives, Reston, Virginia). Together Cerf and Kahn designed TCP/IP, the computer protocol that gave birth to the Internet in the early 1970s, for which they received the U.S. National Medal of Technology in December, 1997, from President William J. Clinton.

In addition to concerns about preserving confidentiality in financial transactions in electronic commerce, verifying the authenticity of transactions, sources, and customers, protecting intellectual property, taxing electronic transactions, and determining liability for illegal content—all of which are already widely recognized and under public debate—

In answer to ARPA's RFQ, a number of responses were submitted. Bolt, Baranek, and Newman in Cambridge, Massachusetts, won the bid. At UCLA, Professor Leonard Kleinrock was tapped to operate a Network Measurement Center, and in September, 1969, the first Interface Message Processor (IMP) was installed, becoming the first node of the ARPAnet. Cerf was then a graduate student working in Kleinrock's group on software for testing and measuring network performance. "The ARPAnet IMPs were essentially packet switches" for directing discrete packets of information through intermediate switches to their ultimate destinations, Cerf recounted. "Today they'd be called routers." The ARPAnet was demon-

Cerf pointed to two other significant open sociotechnical questions.

The first is international agreement about freedom of speech. "Freedom of speech is viewed differently from one country to the next," he observed. "Yet the Internet and the Web are so decentralized, knowing no geopolitical boundaries, that the 'Net is the ultimate tool of free speech. Technological and political attempts to restrict it are not very successful. Take the requirement in China that every Web link must go through a proxy server [a computer that filters every Web search and download]. All it takes is a phone call outside China for a user to land on something that is not a proxy server. Thus, it is no surprise that the Saudi Arabian government has religious concerns about the Internet—because with widespread access, its nation's population would likely be exposed to things they consider immoral. The German government has also expressed concern about the Internet being misused by skinheads or neo-Nazi groups to spread hate speech."

'The Internet is the ultimate tool of free speech.'

The other open issue—which is less about public policy than about socioeconomics—is "that technology and Internet services may create a wider gap between haves and have-nots," Cerf noted. [This issue is discussed in the boxes by Gordon E. Moore in Chapter 4: Computers, p. 94, and by John B. Slaughter in Chapter 12: Preparing Engineers for Tomorrow, p. 284]. "Right now that is true. But throughout history, people with more resources have been better informed than their brethren with fewer resources. The real question is whether that gap will continue to widen or will start to narrow. The jury is out.

"I'm a 'Net optimist," Cerf continued. "First, the 'Net may increase the total amount of information to which all people have access, just as a rising tide raises all boats, even though some may be yachts and others may be dinghies. Even more important, hardware

strated publicly in 1972, after which ARPA continued intense investigation of packet-switching technology for mobile radio and satellite communications.

Meanwhile, other experts were independently working on getting local computers to communicate with one another. Paramount among them was Robert Metcalfe, then at Xerox PARC (Palo Alto Research Center) in Palo Alto, California, who invented the Ethernet local area network in 1973. "Bob [Robert E. Kahn at Bolt, Baranek, and Newman] told me about ARPA's packet radio and packet satellite networks, and we realized [all these networks] would need to be interconnected with the ARPAnet," Cerf explained. "We called it the 'internetting problem,'

and software costs are dropping so fast that the capital cost of being equipped is getting more affordable every year—increasing the opportunities for schools and libraries to offer access to everyone, even if individuals don't have home service.

"What would the price have to drop to for Internet access to be affordable even to low-income households in inner cities?" Cerf speculated. "I daresay that most poor families today have television sets and telephones. Yet, when televisions were first available in the 1950s, they were extremely expensive—$500 to $1,000 in 1950 dollars. Today, half a century later, they can be bought for under $200 in current dollars. So I think that as soon as the cost of dedicated Internet-able appliances—if not full-up personal computers—falls to within the same consumer price range as televisions and telephones, they'll be universally affordable to all families.

"And," Cerf added, "I think that will happen soon—by the end of the decade, if not within the next three or four years. I'm not willing to accept the argument that all low-income families will *permanently* be barred economically from access to the Internet and the Web."

◆

Vinton G. Cerf, who received his IEEE Fellow award in 1988 "for contributions and leadership in the design, development, and application of Internet protocols," is the founding president (1992–95) of the Internet Society. He is a member of the Presidential Information Technology Advisory Committee (PITAC) in the United States and of the Advisory Committee for Telecommunications (ACT) in Ireland. Cerf also serves as technical advisor to production for "Gene Roddenberry's Earth: Final Conflict," the number-one U.S. television show in first-run syndication.

which became the origin of the term 'Internet.' The term was adopted fairly quickly. We had our Protocol for Packet Network Interconnection published in 1974. This paper ultimately spawned the TCP/IP protocols." Internet technology was rolled on top of the ARPAnet, Packet Radio Net, Packet Satellite Net, and Ethernet at the end of 1982.

"The realization that this was going to be a big thing came in steps," said Cerf, who is now senior vice president of Internet architecture and technology at MCI WorldCom Corp., in Reston, Virginia. "I didn't anticipate anything like what has since developed back in '73. In fact, between then and its first deployment,

Tying the world markets together electronically has cost more than one broker a good night's sleep. Extended trading hours, perhaps round-the-clock, will make matters tougher. To keep up with this growth, the North American Security Dealers and Quotations market (NASDAQ) is developing a more powerful monitoring and alert detection system that follows more than 800 trades per second and responds immediately to trading developments. The system can support a $4 billion trading day. *Feature Photo Service and NASDAQ*

I was a carpetbagger with a new gadget to sell, talking it up at every conference." His visionary enthusiasm and hard work paid off. "By '83, everyone in the research community and the DoD was committed to the Internet Protocol"—the TCP/IP (transmission control protocol/Internet protocol) that underlies the workings of the Internet and the World Wide Web today.

"We're just now beginning to accept and recognize that the Internet is going to have impact on all other media—the printing press, the telephone, radio, and television," Cerf reflected. "The Internet is effectively absorbing all their functionality, and adding the ability to have group communications. It also allows the creation in the virtual world of things you couldn't possibly do in the physical world—such as swap meets among 10 million people." [For some of his thoughts about the broader implications of the Internet, see his box "What are key policy and social issues facing the Internet?" p. 4.]

For the twenty-first century, he sees institutions and practices forming that "couldn't exist without the Internet," such as online auction houses, online mer-

WILSON GREATBATCH: What is the secret of happiness in a career?

PHOTO WAS TAKEN BY ROBERT LOUIS, FOR IEEE

"I have a two-minute presentation on success and failure that I always give when I'm invited to lecture at a company or school," said Wilson Greatbatch, inventor of the implantable cardiac pacemaker, for which he has been inducted into the National Inventors Hall of Fame in Akron, Ohio.

"I don't think that the good Lord cares whether you succeed or fail," said Greatbatch. "But I think He does care that you try—and that you try *hard*.

"You shouldn't fear failure, because failures are valuable learning experiences. If everything you do works out well and you have no failures, it means you're not trying hard enough. Moreover, your most abject failure may be part of some grand success in the good Lord's sight that may not even take place in your lifetime.

"Similarly, you shouldn't crave success. If you ask for financial reward, or peer approval—the worst cross that we scientists have to bear—or even gratitude or appreciation for what you do, you're asking to be paid for what you should be doing as a freely-given act of love.

chandisers, global venture capital companies, global software development companies, and others yet unimagined. "They will have a global character, independent of geopolitical location. I think we'll see changes in the political atmosphere because of it, because people with common interests can find expression on the Internet regardless of geopolitics. Already, we're seeing English becoming the second language of choice.

"I think we'll also find a quickening pace of business because business won't go to sleep. The Internet tends to erase time zone differences, because someone somewhere in the world is always awake. Already we're seeing 24-hour order-

"You should do your work because it is a *good* thing to do. Your reward is not in the results, but in the doing.

"If you can get the fear of failure and the craving for success out of your system, it will leave you with a clearer mind to concentrate on the core of the problem in front of you. You can focus on what really needs to be done, free from all the encumbrances the world would so willingly lay on you—and 90 percent of life's stresses will drop away.

"Only then will you find true happiness.

"Our mental hospitals are full of people who couldn't bear failure or bear success," mused Greatbatch. "That will never happen to me, because I just don't care! So I say to students in commencement addresses: go forth and select something you want to do that is a good thing in the Lord's sight. Then study it, work at it, *live it*. Work harder at it than you have ever worked before. Don't fear failure and don't crave success. Just enjoy your total immersion in it.

"And things will work out. The good Lord will smile on your efforts and you'll be left the happiest person in the world."

> **'Don't fear failure and don't crave success.'**

Wilson Greatbatch, who received his Fellow award in 1971 "for vital contributions to biomedical engineering," is an adjunct professor at Cornell University, State University of New York at Buffalo, and Houghton College. Although nominally retired since 1990, Greatbatch's current interests include developing power for implantable pumps for artificial hearts, researching a genetic cure for inherited polycystic kidney disease, and investigating alternative sources of energy.

taking and an expansion of the hours of the world's major stock exchanges. All this poses a pretty tough problem for people who feel they have to keep up with everything—we may all end up suffering from global sleep deprivation. And people are already wondering how we can deal with a surfeit of e-mail and communications." [For a further discussion of that very issue, see the box "When is unlimited information effectively no information at all?" by Rod C. Alferness, in Chapter 5: Communications, p. 110.]

Although the rise of the Internet, videoconferencing, and other sophisticated interactive communications has commonly been cited as a trend that will ultimately decrease the need for travel, Cerf is not so sure. In fact, he said, "the Internet has the funny effect of *increasing* the amount of travel—people using it discover places to go and people they want to meet—so the travel industry may ultimately benefit."

He also does not foresee the paperless office. "Paper won't go away, at least not for a while," Cerf declared. "There are occasions, for example, that I need to see many more pages at the same time than can possibly be displayed even on a very large screen. I don't see paper disappearing until my whole desk can become a display device."

He also feels that electronic books are the wave of the future, but probably will not catch on widely until the technology for reading electronic material becomes as convenient as opening a conventional book. For that to happen, "booting up, logging on, and dialing in must become inconveniences of the past, because the machine is always on and always connected" and battery life is functionally indefinite. Electronic publications will really come into their own when their editors and publishers further embrace the fact that the medium allows them to "convey sound, moving images, spreadsheets, and other types of illustrations that are not achievable on paper."

Technology Saving Lives

"I've never had a government grant," stated Wilson Greatbatch, holder of more than 140 U.S. and foreign patents ranging from tritium-powered batteries to engines that run on alcohol to clones of plants and trees for use as a biomass source of fuel. But the patent for which he is most famous is that for the first successful implantable cardiac pacemaker.

A pacemaker is a device that tells a heart when to beat. Normally, a small electrical signal of a few millivolts is generated by a nerve center in the heart's upper chambers (the atria), telling the heart's two lower chambers (the ventricles) to contract and pump blood throughout the body. But a heart attack or disease can cause an ailment called heart block, in which those natural electrical impulses fail to reach the ventricles in adequate quantity. The patient is left with irregular heartbeats that can cause shortness of breath and perhaps even unconsciousness or death. A pacemaker, by restoring those electrical impulses, keeps the heart beating strongly and regularly, allowing a patient to live out a normal life expectancy.

"I designed the pacemaker on about $2,000 capital using the barn in the back of my house. The barn had very little overhead—about seven feet," he quipped. The reason he went the lone inventor route is that the company for which he then worked was reluctant to bet its future on such a risky device. This was, after all, the

Not the Bionic Man, but he's feeling close to it. The Left Ventricular Assist Device, developed by Thermo-Electron, allows patients to be discharged to live at home while awaiting a heart transplant. Forty-three-year-old Jeff Abner thought his fishing days were over when doctors at Jewish Hospital in Louisville, Kentucky told him that he was dying of congestive heart failure. Now, more than two years after receiving Baxter's Novacor implantable heart-assist device, Jeff regularly fishes with his son near their home in Beaver Dam, Kentucky.
PR NewsFoto, Thermo-Electron Inc., and Baxter Novacor

late 1950s and Greatbatch's prototype had so far kept a dog alive only four months. But Greatbatch had his convictions and his dedication. [For insight into his motivations in life, see his box "What is the secret of happiness in a career?" on p. 8].

Although several groups of investigators in the U.S. and Europe had succeeded in designing pacemakers worn outside the body, they were large and cumbersome and plagued by perpetual infections around the sites where the wires entered

JACK S. KILBY: What are some of technology's unanticipated consequences?

TEXTS INSTRUMENTS

"Over the last decade, there's been a tremendous increase in the ability of people to communicate any time, any way, anywhere," said Jack S, Kilby, inventor of the monolithic integrated circuit, for which he was inducted into the National Inventors Hall of Fame in Akron, Ohio, in 1982. "But we've not yet seen enough of it to understand what the real social implications are.

"Any major change has both up sides and down sides," he continued. "The Internet seems completely innocent, but people have found ways to use it for evil purposes—such as guys hitting on young girls and luring them away from home, as recently happened

the patient's skin to stimulate the heart. Greatbatch wanted to make an entire device—including the power supply—that could reside inside the body.

During his two years of self-imposed barn exile, Greatbatch fashioned 50 pacemakers by hand, using various combinations of transistors (which were then relatively new) and batteries (whose zinc-mercury technology then limited them to a lifetime of about two years). In 1960, William C. Chardack, chief of surgery at the Veterans Administration Hospital in Buffalo, New York, and his associates implanted Greatbatch's pacemakers into 10 human patients using heart electrodes of Chardack's design. All had complete heart blocks and perhaps only a 50 per-

around here. It's a minor proportion, but it's real.

"But I don't think there should be centralized controls," he said. "I think everyone has to be aware of the possibility that any change will have its good and bad effects. For example, newspapers are now filled with articles on the hazards of automobile air bags, which were viewed as an unmixed blessing 10 years ago.

'Any major change has both up sides and down sides.'

"Every technology has its unanticipated consequences. And many of them are good. The personal computer is one," Kilby noted. "When mainframe or mini computers cost a million dollars apiece 25 years ago, anyone with one would have shot you if you'd wanted to use it to write a letter! Who then would have thought that 25 years later one of the computer's main uses would be word processing, and that PCs would be sold in retail stores for under $1,000 or even under $500?"

◆

Jack S. Kilby, who received his IEEE Fellow award in 1966 "for contributions to the field of integrated circuits through basic concepts, inventions and development," was responsible for all integrated circuit development at Texas Instruments in Dallas from 1958 to 1970. He is the holder of more than 50 U.S. patents for work on ICs, including the first patent for a hand-held calculator (the Canon four-function Pocketronic of 1972) and the first patent on a semiconductor thermal printer.

---

cent chance of living another year. Although one died in 24 months, the longest-lived one went on for another 30 years, his heart paced the whole time.

In 1961, Greatbatch signed a license for the pacemaker with Medtronic Inc., whose Chardack-Greatbatch Pacemaker would dominate the field for the next decade. The rest, as they say, is history.

But Greatbatch never ceased his quest in devoting his engineering creativity to improve human life. Troubled by the short battery life, he worked on improving lithium batteries and became the first to incorporate them in a fully implantable biomedical device. Today, lithium batteries for pacemakers have a life of 6 to

10 years, and Greatbatch Ltd. in Clarence, New York, the company he founded to manufacture them, still sells or licenses more than 90 percent of the world's pacemaker batteries.

"The pacemaker is getting to be a fairly mature technology," Greatbatch remarked, an understatement for a technology that is now implanted into 600,000 patients worldwide each year.

In his view, the current biomedical engineering frontier for cardiac patients is an implantable left-ventricle assist device (LVAD)—"essentially, half an artificial heart," Greatbatch said. "An LVAD assists the left ventricle, the heart's main pumping chamber." In the U.S. alone, some 20,000 people a year need heart transplants while there are hearts enough for only about 2,000. Hence Greatbatch's sense of urgency about developing the LVAD for permanent implantation.

But the technical challenge is orders of magnitude more difficult than that which faced the pacemaker. "The pacemaker just emits a tiny electrical signal that says 'beat,' and the natural heart does all the work," Greatbatch explained. "The LVAD, however, is a mechanical pump doing actual physical work. The same battery that can power a pacemaker for 20 years could drive an LVAD only 20 minutes." This power requirement means patients wear bulky belts of rechargeable batteries with a radio transmitter to continually recharge the implanted batteries. As a result, LVADs are usually used only temporarily for patients waiting for transplants of natural human hearts.

Moreover, most LVADs are not yet completely implantable because they require a drain tube through the skin to drain bodily fluids that accumulate behind the pump's diaphragm. In addition, "artificial hearts so far have really not achieved complete biocompatibility between the pump materials and the blood itself," Greatbatch said. "One possibility is to treat the surface of the synthetic rubber diaphragm with a material that encourages tissue growth, so the blood eventually sees it as a vessel."

Limitations aside, Greatbatch feels that the major biomedical accomplishments of the twenty-first century will include "a fully functional implantable LVAD that will last the life of the patient." Also, he noted, "people are working on artificial urinary bladders and other artificial internal organs. I can't even begin to guess the progress that might be made."

After the Integrated Circuit . . . What?

"Did I envision the revolution that would be brought about by the invention of the integrated circuit?" asked Jack S. Kilby, its inventor in 1958. He snorted. "I

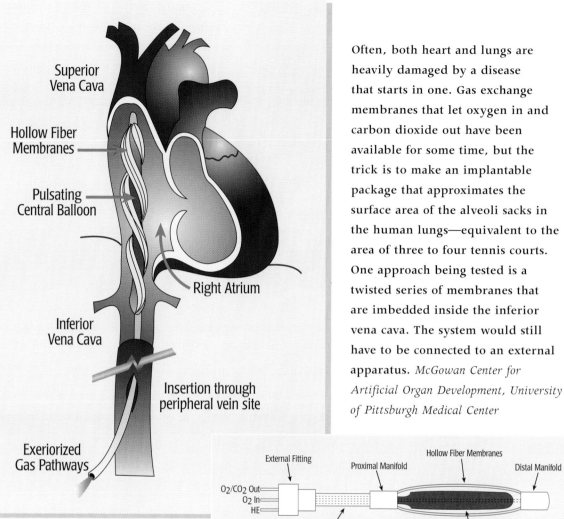

Superior
Vena Cava

Hollow Fiber
Membranes

Pulsating
Central Balloon

Right Atrium

Inferior
Vena Cava

Insertion through
peripheral vein site

Exeriorized
Gas Pathways

Often, both heart and lungs are heavily damaged by a disease that starts in one. Gas exchange membranes that let oxygen in and carbon dioxide out have been available for some time, but the trick is to make an implantable package that approximates the surface area of the alveoli sacks in the human lungs—equivalent to the area of three to four tennis courts. One approach being tested is a twisted series of membranes that are imbedded inside the inferior vena cava. The system would still have to be connected to an external apparatus. *McGowan Center for Artificial Organ Development, University of Pittsburgh Medical Center*

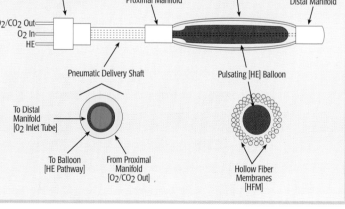

External Fitting

Proximal Manifold

Hollow Fiber Membranes

Distal Manifold

O_2/CO_2 Out
O_2 In
HE

Pneumatic Delivery Shaft

Pulsating [HE] Balloon

To Distal
Manifold
[O_2 Inlet Tube]

To Balloon
[HE Pathway]

From Proximal
Manifold
[O_2/CO_2 Out]

Hollow Fiber
Membranes
[HFM]

certainly did not. And I don't think anyone else could either. Who could have ever predicted that it would set off a decrease in the cost of electronic functions of fully 10 million to one? I don't think anyone had any basis for extrapolating such change. The magnitude of that change is as if full-sized cars now cost U.S. $100 apiece." [For more of Kilby's thoughts about the inherent unpredictability of technology, see his box "What are some of technology's unanticipated consequences?" p. 12.]

The monolithic integrated circuit (IC) is the guts of every "microchip" in today's high-tech devices. In concept, it is elegantly simple and powerful. Instead of building a circuit out of individual discrete components (resistors, capacitors, transistors, inductors) that are individually wired or soldered together, a circuit is

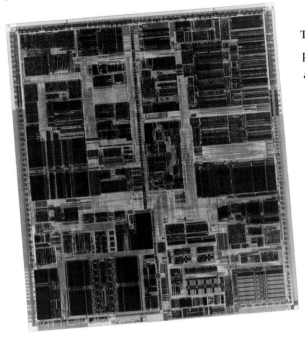

The inside view of an Intel P6 central processor could be mistaken for an aerial view of a factory. The two have much in common since both take in a product (raw data) and a customer's orders (programming) and turn them into a finished product (images, instructions, or letters on this page). Like conventional factories, the electronics revolution has been marked by smaller, more efficient processors to meet demands for great productivity.
Intel Corp.

built up in layers by lithographic techniques inspired by the multilayered art of color silk-screen printing. Starting with a single substrate or wafer of silicon, different areas are masked, implanted with impurity atoms to give them desired electronic properties, etched away, and grown with more silicon to yield structures that behave as independent components.

Moreover, it was Kilby's thought that the processes by which transistors were made could be used for fabricating other types of electronic components. And if one could do that, then it would be a relatively simple matter to fabricate them on the same piece of material and to interconnect them. The smaller the transistors and conductor lines could be made, the faster and denser the chip could be—limited primarily by lithographic and manufacturing techniques.

And the rest is four decades of exponential growth in speed and capacity, and exponential decline in cost.

What is now particularly amazing to Kilby is that "people feel the IC has a productive future for at least another 15 years, as is projected by the semiconductor industry road map," he said. "The basic concept of the IC is more than 40 years old now. Forty years is as long as the vacuum tube lasted—and in the fast-moving world of high technology, that's a long time!"

He wonders now if there are inherent limits to the concept of the integrated circuit. "Much progress has been made using finer and finer geometries," he said, referring to the widths of the lines that could be lithographed onto the chips to make them denser and faster. "Today we're down to 0.25 micrometer [about $1/400$ the width of a human hair], and people have made experimental chips with line widths as narrow as 0.1 micrometer.

"After that, it's not too clear what may happen, as existing transistor structures don't work very well when they're so small," he explained. Among other things, such small devices become sensitive to the fact that the impurity atoms with which different structures are doped to give them specific electronic properties "aren't distributed completely uniformly," he said, so they don't work as expected. "But people have projected an end to current IC design off and on for quite a while, and so far whenever we've gotten close to a supposed limit, we've worked around it," he noted.

"Before long, though, I suspect that something completely new will appear on the horizon." When asked if he had any idea what, he firmly replied: "Not the slightest!"

Technology in the Service of Humankind

"Dramatic changes come when multiple technologies merge to meet human needs," said Arno A. Penzias, former vice president of research at AT&T Bell Laboratories. From the vantage point of his various managerial positions in encouraging research over several decades at Bell Labs (later Lucent Technologies), he came to the realization that combinations of existing technologies to meet new needs are what will be "the big driver of the next century." [For Penzias' reflections on the significance of his personal Nobel Prize-winning cosmological research, see his box "What constitutes a good scientific theory?" p. 18.]

The World Wide Web, for example, "happened because for the first time

Zenith engineer Jeanette Brown tests circuit modules of the Zenith VSB prototype receiver used in high-definition TV modulation equipment at the company's research center in Glenview, Illinois. *Zenith Corp.*

ARNO A. PENZIAS: What constitutes a good scientific theory?

BELL LABS

"Fame for having been involved in fundamental advances in science often puts you in the position of having to defend or explain science—of having to answer such questions as 'how can you *prove* that?' That gets down to examining basic assumptions of what science is or isn't," mused Arno A. Penzias. Penzias and Robert W. Wilson were awarded the 1978 Nobel Prize in Physics for their 1965 observational discovery that the universe is bathed in microwave background radiation at a temperature of 3 K. Their observation was widely acclaimed as crucial evidence supporting George Gamow's Big Bang theory of the origin of the universe. The Big Bang theory, in Penzias' words, postulated "that the universe was created out of nothing with a positive energy, that it seems to expand forever, and that the laws of physics applied immediately after the moment of creation."

"I originally came to science thinking that theories are proven or disproven," Penzias continued. "In actuality, they are accepted or abandoned. Scientists can't prove theories in absolute terms—'proof' comes down to practical experience.

personal computers had enough memory and processing power that people could put bit-mapped displays in place so users could look at images instead of text. At the same time, communications networking was also good enough to support the needed data rates." And, of course, the cost of all the technology was cheap enough to be widespread throughout the populace.

"Some of it looks obvious in retrospect, and also looks simple in prospect," Penzias said. "But picking the right combination takes intuition. It has to be a combination that meets a market need. That's the hard part: seeing what it is that people still need that they don't have.

"When scientists describe science to the public, though, we tend to present our stuff as truth. But we tend to forget underlying assumptions that are basically unprovable—such as, that the simplest theory is the right one. There's more of a human element in science than the public usually knows. Science is not so unreliable that any crazy idea is to be admitted—but it's a thin line, and a lot shakier than supporters of the scientific method would have you believe.

"My personal perspective is that when we discovered the microwave background radiation in the 1960s, cosmology became an experimental science for the first time. That observation implied that the universe is really knowable and can be subjected to laboratory tests in a way that was not possible before. The fact that we have so much observational evidence about cosmology means we can no longer say—as we could before the 1960s—that we have solved the problem but we just can't test the solution."

The very existence of so much observational evidence, however, poses a conundrum that has "driven cosmologists to a state of uncomfortable awkwardness," Penzias observed. Although the existence of the 3 K microwave background radiation was a central prediction of the Big Bang theory, the Big Bang theory itself begs disturbing questions: What existed before the Big Bang? What caused the Big Bang? *Why* did the universe come into being? In short, "you're left with a mystery that goes beyond physics," Penzias said. "All physics does is describe one thing in terms of another. It doesn't *explain* things."

So, for the past 35 years, many cosmologists have been proposing alternative theories of "a universe that needs no explanation," Penzias said. "A universe that needs no explanation is one that's always been here, or one that needs no net energy, or one in which the observed excess of matter over antimatter is a random local variation." There are, for example, quantum cosmological variants on an older theory that the universe is actually oscillating—that the Big Bang for which Penzias co-discovered observational evidence is only the most recent of an infinite series of Big Bang explosions, because the

"Now, to many engineers, combining existing technologies doesn't look 'techie' enough. They undervalue marketing—the ability to pick the right combination—because they confuse it with sales. But engineers have to remember that a big part of all technological advance is understanding market needs, because otherwise they'll put the wrong things together," he said, and perhaps end up with a market flop. What's most tricky about marketing, he's found, is that "customers themselves often don't know what they want."

But engineers can gain a sense of potential markets by listening to people and observing them closely—and even watching and observing themselves, Pen-

universe keeps collapsing back onto itself. In other words, the universe is closed, not open.

"But each of those alternative theories has observational consequences that have to be explained away," Penzias noted, "such as postulating subatomic particles for which there are no other data." All the theories for closed cosmological models, for example, require that the universe be much more massive than it is observed to be. "The 'missing mass' is known not to exist," Penzias said. "There are at least four ways of measuring the mass of the universe, and none is consistent with a theory of a closed universe. Moreover, you need inflation [a theoretically postulated moment of ultra-rapid expansion just after the initial instant of creation] of the universe to account for the observed excess of matter over antimatter.

"You can always make a theory more complicated to fit the facts. And quantum cosmologists, in their search for a universe that requires no explanation, have been driven to extraordinary complexity—even reviving Einstein's cosmological constant [a repulsive force to hold the galaxies apart in a static universe] to diddle the rate of the universe's expansion through antigravity.

"Cosmologist Steve Weinberg said, 'The more I look at the universe, the more meaningless it becomes.' But quantum cosmologists have had to go through all sorts of contortions in their theories to keep it meaningless. And just because a complicated theory can be made to fit the data doesn't mean it's correct. The theories are so contrived, in fact, that workers in any field other than cosmology wouldn't look at a theory whose assumptions are so complicated. And the experimental evidence is against many of them, besides," Penzias said.

"In the past, we've always been able to demystify natural phenomena with scientific theory. But the very complexity to which quantum cosmologies are driven makes me, at least, believe that we are still left with a universe whose origins are a total mys-

zias advised. "What are the things that annoy you?" he asked. "What are the things that make you say, 'I wish they would . . .' Then think about becoming the 'they.'"

To take just one example, in Penzias' view, "transportation will be a huge win" in this type of combinatorial thinking. "I am firmly convinced that within 10 or 20 years, there will be an enormous renaissance in public transportation. There is no place for continued growth in private automobiles." Even in U.S. cities such as Boston, Los Angeles, or New York City, where rush-hour driving is an exercise in tedium and frustration, "we are *spoiled* compared to São Paolo or

tery. We are still left with the mystery that the universe has more matter than antimatter everywhere, and that there is less mass than is necessary to bind it together—all of which implies a single expansion that is destined to go on forever. Much as cosmologists have tried to paper it over with their baroque theories, the Big Bang model is still the simplest explanation of the observations.

'Scientists can't prove theories in absolute terms—"proof" comes down to practical experience.'

"Moreover, no theory yet explains how or why the universe came into being. The universe seems to have an inherently unknowable—I won't say Biblical—part to it. So we certainly can't say that everything's been done and there's no room left for wonder.

"So we enter the twenty-first century with the same picture that George Gamow would have understood half a century ago," Penzias mused. "And we are still left with the mystery of existence.

"That, to me, is remarkable."

◆

Arno A. Penzias is a Venture Partner at New Enterprise Associates in Menlo Park, California. His scientific career began when he joined Bell Laboratories in 1961, and culminated in a series of managerial positions in Bell Labs' research organization, including vice president of research. He is the author of Ideas and Information: Managing in a High-Tech World *(New York, N.Y.: W.W. Norton, 1989) and* Digital Harmony: Business, Technology and Life After Paperwork *(New York, N.Y.: HarperCollins, 1995).*

Bangkok," Penzias asserted. "We don't *know* what traffic jams are—or how bad air pollution can be, as it is in cities like Beijing."

In Penzias' view, limousine-quality public transportation could increase its attractiveness enough to draw significant ridership—complete with home pagers that would alert each regular passenger when the bus is about to reach his or her stop. Global Positioning System (GPS) input to routing maps could alert the driver to traffic jams ahead and could plot out a suggested alternative route—a technology already commercially available from at least one high-end auto maker. It would even be possible for people to have a business meeting on the way to work, he

An ambulance en route to a hospital in San Antonio, Texas, uses the TransGuide advanced traffic management system to avoid traffic problems in life-threatening situations. TransGuide's LifeLink program allows two-way video, audio and data teleconferencing between paramedics and emergency room doctors. TransGuide uses fiber optics, sensors, and video cameras to detect changes in traffic flow. Messages about road conditions, exit closures, and accidents are then displayed on signs located on highways throughout San Antonio.
PR NewsFoto and TransGuide

noted. "These are simple things we could do today," he said. Ultimately, a local cab company might "use as many MIPS [million instructions per second] as an international airline."

Forefronts for Lasers

"Laser technology is still fairly new—in its adolescence," observed Charles H. Townes, co-winner of the 1964 Nobel Prize in Physics for the fundamental technology underlying the invention of the laser. "Lasers marry the masterly techniques of electronic control with the all-pervasive field of optics," said Townes, to create "technical possibilities in a wide variety of fields that we couldn't imagine before." [For some of Townes' thoughts about the unpredictability of technological applications and the resultant necessity for long-term research, see his box "How can we be wiser about decisions of what to pursue in science and technology?" p. 24.]

In Townes' opinion, lasers promise to be great scientific tools in the early twenty-first century for the exploration of both chemistry and biology.

"Most interactions in physics are with surfaces," Townes explained. "But surfaces are not well understood, compared to what we know about bulk materials, gases, and liquids." Surfaces are fundamentally different from bulk materials because their atoms and molecules are anchored on only one side instead of on all sides. That sometimes makes them unusually reactive. As a simple example, water on an iron surface readily oxidizes the metal to form rust. Molecules that attach to surfaces are vital in manufacturing and to catalysis: "the catalytic converter in automobiles cleans up the car exhaust through surface interactions," Townes said. "People know about surfaces empirically and use their reactive effects, but a detailed understanding is skimpy."

Lasers, however, may change that. When combined with both nonlinear optics and a technique known as Raman spectroscopy—the mapping of the wavelengths of light scattered from a material—lasers can help differentiate light scattered from surface molecules from that scattered from bulk molecules deeper inside the material. "The surface molecules oscillate differently and respond differently to intense light, so produce different spectra," Townes said. In his view, lasers will open up further study because they allow "more sensitivity and specificity," he said.

In addition, many materials are normally attracted toward intense light because of their index of refraction (ratio of the speed of light in a vacuum to the speed of light in the material). Since a laser beam is far more intense at its center than at its edges, this nonlinearity means that a laser beam can actually act as tweezers to move or pick up small objects.

According to Townes, all you need is to float a small object, such as a living cell, in a medium (such as water) that has a slightly smaller index of refraction. When a low-power laser beam is focused into the water, the object will naturally be attracted to the focus of the beam "almost like falling into a gravitational well," because the molecule's energy inside the bright field is less than that at the edges. "Thus, you can pick up small objects with the laser beam," Townes said. This is of great interest in biological studies, because "you can pick up single cells, or move parts of a cell around, without injuring them," or "you can stretch out the double helix of a DNA molecule to look at it better."

Short pulses from lasers can also drill tiny holes through metal, diamond, or even human flesh by quick evaporation that does not heat or chemically alter the surrounding atomic layers—giving rise to unusually precise and clean holes that are useful both in metallurgy and fast-healing surgery, he said. Moreover, pressure waves from laser pulses can quickly flash-heat the surfaces of materials,

CHARLES H. TOWNES: How can we be wiser about decisions of what to pursue in science and technology?

"In the early days of the laser, people kidded me that it was a solution looking for a problem," remarked Charles H. Townes, professor at the University of California at Berkeley. In 1964, Townes shared the Nobel Prize in Physics with Soviet scientists Alexander Prokhorov and Nicolai Basov for fundamental work in quantum electronics, which led to the construction of oscillators and amplifiers based on the maser-laser principle. "'What can it be used for?' they asked. Now, of course, we know that was a limited, short-sighted approach. But back then, nobody had any grasp of what the laser could do.

"Very basic discoveries can come out of applied work, and we know many applied things that have come from basic work. It was clear to Dr. [Arthur L.] Schawlow [co-inventor of the laser] and me that lasers would have many

allowing a thin surface layer to be annealed (hardened) against scratching without affecting the flexibility of the underlying structure (solving the long-perplexing problem in traditional bulk annealing processes of how to prevent a material from becoming brittle).

Predictions: Interplanetary Internet

Vint Cerf is now working with the National Aeronautics and Space Administration's Jet Propulsion Laboratory in Pasadena, California, and with the Department of Defense's Advanced Research Projects Agency (now called DARPA) to

applications because they emitted such pure and intense radiation. As I was interested in spectroscopy, for example, I could foresee a lot of interesting scientific uses. We could also foresee useful technological applications, such as communications or industrial applications requiring an intense concentration of power. In fact, we used to argue broadly that since the laser marries electronics and optics, it was bound to touch on a wide range of technologies.

"But we did not think of fiber optics, nor did we imagine the laser's usefulness in biological and medical applications, such as reattaching detached retinas. Although we recognized some nonlinear effects, we had no idea of the richness of the field of nonlinear optics and how useful it would be in opening up a wide range of new phenomena. Those include new spectroscopic responses of materials and the production of solitons," he said, referring to self-sustaining pulses that travel in optical fibers.

"Thirty years ago we knew we couldn't predict everything. But we also knew we needed to explore what the laser could give us. And indeed, laser technology has expanded even more than what Dr. Schawlow and I could dream in the 1960s. And now I have every confidence its usefulness will continue to grow in the twenty-first century."

In Townes' view, the unexpected richness of the laser's history in science and engineer-

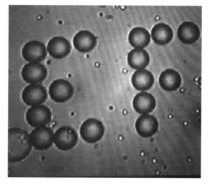

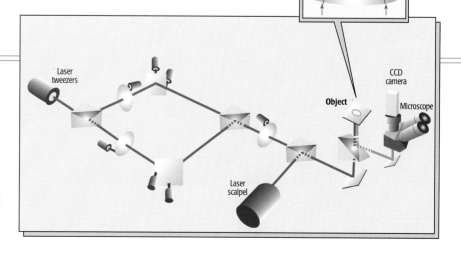

Several institutions are developing lasers as optical tweezers and scalpels because of their potential for manipulating individual cells or even proteins. This image at upper left shows the result after manipulation of 10 micrometer spheres using an optical tweezers depicted in the schematic under a 60x microscope (Experiments usually are performed under 100x). *University of Umea*

> 'As a society, we must be sure we don't focus all efforts just on things we are *sure* will pay off economically.'

ing serves as a vital lesson: "As a society, we must be sure we don't focus all efforts just on things we are *sure* will pay off economically. We need to devote some resources to exploring things that may revolutionize our understanding. We must continually emphasize that, and take the risk.

"Too often today you hear the argument that 'we have only limited dollars, so let's spend them on something useful.' Businesses especially have that viewpoint, because they must report their earnings every three months—so they feel they have to look very short-term and get results quickly. Moreover, since a company executive is seldom in place more than 10 years, he usually wants to do something during his tenure that looks good for him.

"But the long-term scientific and technological needs of our nation require a longer view than the next quarter or even the next ten years. So the federal government has to act as an insurance agent for financing longer-term work in universities and government laboratories. Massachusetts Institute of Technology economist Robert Solow [winner of the Nobel Prize in Economics in 1987] has tried to quantify the payoff of research. In the United States, for every dollar of products sold, less than a penny is invested in R&D— yet R&D has much more than a 1-percent effect on the final product. So economically speaking, investing more in scientific research is sensible."

Appreciating the value of a long-term perspective and providing wise support, however, means that "politicians and the population at large need to understand how the technological developments of today stem from developments of the past," Townes ob-

extend the Internet off the earth to Mars. "This isn't speculative," Cerf declared. "We're engineering this for several Mars missions that will begin launching in 2003. By 2008, we'll have a two-planet Internet."

The immediate use for a Martian Internet will be for effective communications among the rovers and other assets on the Martian ground, six satellites in low Martian orbit, one satellite in synchronous orbit (an orbit in which it keeps pace with Mars' rotation on its axis, so it stays above one geographical area of Mars during the planet's 24-hour, 37-minute day), and the earth. Images, sounds, and measurements from the planet's surface could flow back to Earth. Earthbound

served. Noting that the fundamental laws of electromagnetism that underlie most of the twentieth century's high technology were first articulated more than a century ago, he said: "Legislators and citizens don't have enough understanding of science and how discoveries come about. We need more people who are educated in science and engineering going into public affairs.

"In reality, the things we *don't* know about now will probably be still more useful than the things we think we can foresee," Townes said thoughtfully. "Most of the truly important developments in the twentieth century have come as a surprise. So in the twenty-first century, we have to remain open to new things. Yes, some esoteric explorations *won't* produce economic results. But others will be revolutionary. And at the outset, you do not necessarily know which will be which."

'Most of the truly important developments in the twentieth century have come as a surprise.'

◆

Charles H. Townes, who received his Fellow award from the Institute of Radio Engineers (one of the two predecessor societies of the IEEE) in 1962 "for fundamental contributions to the maser," is University Professor of Physics at the University of California at Berkeley. During World War II, he designed radar bombing systems and computing devices. He also was a pioneer observer of microwave observations of interstellar gas clouds, leading to the first discovery of molecules in space—ammonia and water.

scientists could respond and direct the artificially intelligent rovers' movements and scientific experiments in under an hour—a dramatic contrast to the weeks required in the 1970s to direct the Viking landers to scoop up Martian soil and chemically analyze it.

The ultimate use for an interplanetary Internet, though, will be to "make the Internet accessible and workable for people mining asteroids or living in research outposts on the Moon or in orbit around Mars," Cerf said. "We can even anticipate interplanetary colonization by the end of the twenty-first century."

Why work on protocols and interplanetary e-mail addresses so far in advance?

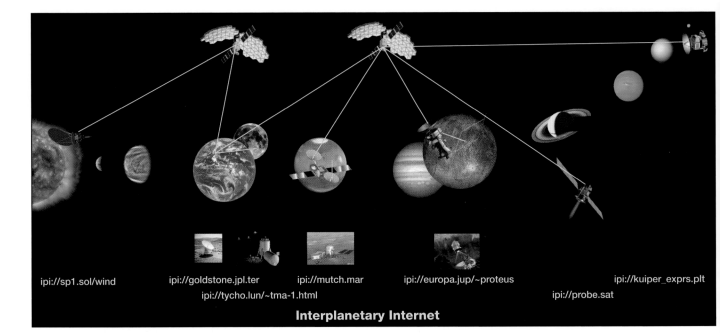

ipi://sp1.sol/wind

ipi://goldstone.jpl.ter

ipi://mutch.mar

ipi://europa.jup/~proteus

ipi://kuiper_exprs.plt

ipi://tycho.lun/~tma-1.html

ipi://probe.sat

Interplanetary Internet

The Interplanetary Internet would coordinate access to data to and from space probes and manned space missions. The Earth node (with a fake address, as are the others) would comprise 70-meter-diameter antennas of the Deep Space Network communicating with counterparts in solar orbits inclined to Earth's orbit so they would always be visible, even when on the other side of the Sun. Manned and robotic missions to Mars and other worlds would link to the orbiting DSN stations via communications satellites orbiting their own worlds. Links between satellites eventually will use laser communications to improve bandwidth and data integrity. *Dave Dooling with art from NASA, NASDA, the Mars Society, and Hughes Aircraft*

"I'm very conscious that it took more than 20 years to get where we are now in today's Internet—getting the protocols and the applications written on top of them," Cerf said. Plus, an interplanetary Internet faces challenges that are lesser issues here on the earth—such as the universe's unbreakable speed limit, the speed of light. Even at 300,000 kilometers per second, signals still require anywhere from 10 to 40 *minutes* to make a round trip between Earth and Mars, depending on the relative positions of the planets in their orbits—making exceptionally long intervals for accessing Web pages across interplanetary distances.

But Cerf and his JPL colleagues have the backing of $20 million extra seed money from the U.S. Office of Management and Budget to engineer a working system. "I'm very excited by the whole concept," Cerf exclaimed. "And the core team deeply cares about it. It's taking on a life of its own. And somewhere around 2020, we'll see an awakening of true commercialization of space." And the Internet will be there and waiting with "interplanetary gateways and a backbone across the solar system." ◇

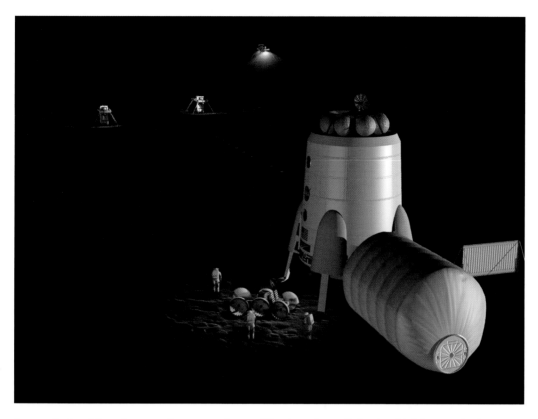

Instead of gold and gems, lunar settlers will mine the soil for ice hidden in dark corners of the lunar poles, and for helium 3 baked into the soil by the solar wind. Ice can be electrolyzed into oxygen and hydrogen for use as rocket propellant; helium 3 could fuel a fusion-based economy when fusion power becomes practical. *John Frassnito & Associates for NASA*

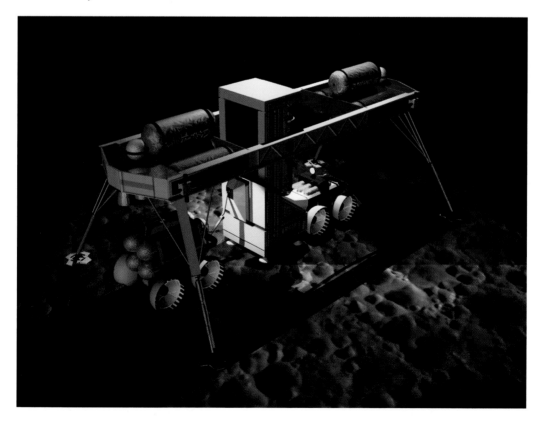

CHAPTER 2

STRUCTURES AND DEVICES

OFTEN, THE SECRETS TO TECHNOLOGY'S SUCCESS are developments inside the "black box" or behind the walls or under the ground or in the air—the technological infrastructure that is itself invisible to consumers, but has profound effects on the products available and even on our way of life.

In this chapter, four experts highlight just a few of the "invisible" infrastructure technologies that may bring revolutionary changes to consumers in the twenty-first century. John G. Kassakian, professor of electrical engineering and director of the Laboratory for Electromagnetic and Electronic Systems at the Massachusetts Institute of Technology in Cambridge, Massachusetts, discusses how humble power electronics will revolutionize not only the engine performance and suspension but also the styling of automobiles.

Also focused on cars—especially on the promise of millimeter and sub-millimeter-wave electronics for vehicular radar systems—is Roger D. Pollard. He holds the Hewlett-Packard Chair in High Frequency Measurements at the School of Elec-

Left: Advanced computers in the twenty-first century are expected to be based on light rather than electricity. Early steps in that direction are being taken with non-linear optics (NLO) that process several data streams simultaneously and work at the speed of light. Don Frazier, a materials scientist at NASA's Marshall Space Flight Center in Huntsville, Alabama, is working on processes for forming NLO thin films in orbit where weightlessness virtually eliminates defects. These systems can be used for automated fingerprinting, photographic scanning and the development of sophisticated artificial intelligence systems that can learn and evolve. *NASA/Marshall Space Flight Center*

tronic and Electrical Engineering at England's University of Leeds, and is deputy director of the University's Institute of Microwaves and Photonics.

Donald R. Scifres, chairman of the board and CEO of SDL Inc., San Jose, California, describes how semiconductor lasers are now becoming so powerful—and have so many major advantages—that they are poised to replace almost every other type of laser in heavy industry for cutting, welding, and materials processing. And Ralph C. Merkle, a research scientist at Xerox Palo Alto Research Center (PARC) in California, and co-recipient (with Stephen Walch) of the Foresight Institute's 1998 Feynman Prize in Nanotechnology for Theoretical Work, contends that such heavy industrial manufacturing will eventually be a relic of the past—replaced by the atomic precision of molecular manufacturing.

Behind the Scenes I: Power Electronics

"The most ubiquitous consumer use of power electronics is in light dimmers— the first really large-scale consumer application of solid-state power electronics," said MIT's Kassakian. To be sure, industry has been using power electronics since

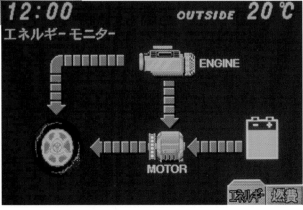

The inverter on the Toyota Prius hybrid vehicle is one of the earliest examples of power electronics to be employed in automobiles. The inverter changes direct current from the battery into alternating current for the electric motor, and alternating current from the motor and generator into direct current for the battery. The inverter features an intelligent power module for enhanced dependability. *Toyota*

vacuum tube days for controlling industrial processes, such as electroplating, electrochemistry, and high-voltage direct-current power transmission. "But it wasn't until solid-state electronics came along that the breadth of applications exploded," he said.

Unlike signal electronics—such as the microprocessor or computer chips that route and process information—the purpose of power electronics is to route and process energy. "But information is invariably used to control energy, so power electronics acts as the interface between information and energy," Kassakian explained.

Power electronics often may control the speed of electric motors, such as the ones that drive the cascaded paper rollers in a paper mill or on a printing press. "All the rollers have to go at the same speed to keep the webs of paper under constant tension, otherwise the paper will rip," Kassakian pointed out. "So power electronics controls are continuously monitoring conditions and using feedback to adjust the speed."

At other times, power electronics may be converting energy from one form to another. One example is converting the stored chemical energy in a battery to mechanical energy for driving the wheels of an accelerating electric vehicle, or converting the kinetic energy from the wheels back into chemical energy to recharge the battery as the vehicle slows to a stop. "Electric vehicles require tremendous power electronics," Kassakian remarked. "In fact, the electronics for controlling the motor cost five to 10 times more than the motor itself."

Just as the goal in signal electronics is to get the density of information processed as high as possible, the goal in power electronics is to increase the density of *power* processed. And just as signal-electronics engineers aim to reduce noise (so as to reduce signal strength and thereby put the maximum amount of information into the minimum area without interference), power-electronics engineers aim to increase the efficiency of their chips in dissipating heat. Because heat dissipation from high voltages and currents is, in fact, a big limiting factor for power electronics, the chips are physically much larger than more familiar computer chips—resulting in electronic components weighing up to a kilogram (a couple of pounds), Kassakian noted.

"One problem with trying to explain the importance of power electronics is that their application is generally not visible to the ordinary consumer," he noted. But indeed, they are everywhere in consumer applications—from the electronic ballast in compact fluorescent lamps that can now be screwed into standard incandescent light sockets, to the piezoelectric actuators that convert mechanical stress into a current that flashes the light-emitting diodes in the heels of kids' sneakers.

JOHN G. KASSAKIAN: How soon can we free ourselves from fossil fuels?

"Fossil-fuel energy has been both a blessing and a curse for Western society," observed John G. Kassakian, professor of electrical engineering and director of the Laboratory for Electromagnetic and Electronic Systems at the Massachusetts Institute of Technology in Cambridge, Massachusetts. "Sources of fossil-fuel energy such as oil, coal, and natural gas are absolutely essential to its very existence.

"But those sources are also very vulnerable," he pointed out. "Fossil-fuel prices are highly volatile, and depend on international politics in the Middle East. They have been the cause of wars and thousands of deaths. And their vulnerability is costing Western society a very high price in domestic well-being and international policies. We are being held hostage politically and economically.

They are even used in toasting the popular American breakfast cereal Cheerios, thanks to Kassakian himself, who designed an electronic induction cooking system for General Mills to replace the natural-gas burners used for toasting until the oil crisis of the late 1970s.

Power electronics is also crucial in the redesign of the electronic system of private automobiles that will debut in the first decade of the twenty-first century. "The last major change in the car's electrical system was the change [from a 6-volt] to a 12-volt system in 1953," Kassakian said. But cars based on 36-volt batteries are now in prototype. That change, made possible by reliable solid-state

"Moreover, their effects on the environment are not benign," Kassakian continued. [For more discussion of the global environmental impact of the use of fossil fuels, see Chapter 10: The Environment, and the box by Robert A. Bell on p. 58 of Chapter 3: Techniques and Systems] "Yet, as developing countries come into the twenty-first century, there is going to be very rapid industrialization—and thus rapidly growing demand for energy."

In Kassakian's opinion, the only choice is to move to some more readily available alternate source of energy.

But that is easier said than done.

"The problem is, fossil fuels are incredibly energy-dense," he explained. "When you drive up to the pump and put the nozzle into the tank and transfer gasoline into your car at three gallons per minute, that is the equivalent of a power of 6 megawatts." (Power, you recall, is the rate at which energy can be supplied.) "Now, in electrical terms,

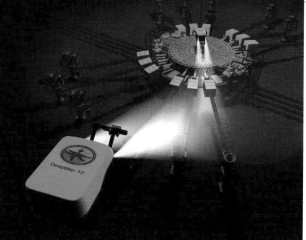

Farther out and deeper is where companies are looking for petroleum offshore as industrial economies consume fossil fuels. This is exemplified by the PennzEnergy Co.'s exploration drilling in the Gulf of Mexico (top) and Texaco's Deepstar plans for undersea wellheads maintained by remotely operated vehicles (right). *PR NewsFoto (top); John Frassnito & Associates (right)*

'When you transfer gasoline into your car at three gallons per minute, that is the equivalent of a power of 6 megawatts ... about the same as the power supplied to 1,000 average American homes.'

6 megawatts is a *lot* of power," he continued. "That's about the same as the power supplied to 1,000 average American homes. There's no way you could charge an electric vehicle at 6 megawatts—the battery would explode. That's why the battery needs to be charged for several hours or overnight rather than for three minutes."

Few alternative sources of energy have the same energy density as fossil fuels, he pointed out. "Solar and wind have such low energy density that in many places they just can't economically justify the expense of the physical structures required to capture the energy." Hydroelectric power is reasonably energy-dense, but large rivers are in relatively short supply and "there are ecological issues to be addressed" in building large dams and flooding whole valleys and their natural habitats, he said. Fusion power has high energy density, but "so far its promises have not been filled—maybe we'll have it in viable form in another hundred years."

Fuel cells, which through a chemical reaction separate ions (electrically charged atoms) to create an electric field from which energy can be extracted, "are great for generating electricity from hydrogen," Kassakian said. "But the problem is, where to get the

power electronics and microprocessors ("people don't realize how much computing power is in a car"), will revolutionize automotive performance, suspension, and even styling.

Instead of mechanically actuated engine valves driven by an overhead camshaft connected by a timing chain to the crankshaft, as found in present-day cars, the valves in twenty-first-century cars will be electromechanical and individually controlled by power electronics. "In today's cars, the timing is fixed, and is a compromise for all engine speeds and loads," Kassakian explained. "But if valves can be controlled individually, you can optimize the timing for every speed you're

hydrogen? Everyone wants to use a reformer that extracts it from gasoline or methanol, but one byproduct of the extraction is carbon monoxide, which degrades the fuel cell if not removed, and thus adds more complexity and expense. So far fuel cells are too expensive for commercial manufacturing," although he does feel that in the twenty-first century, fuel cells will ultimately be the most viable solution to the pollution problem of exhaust emissions in private automobiles.

In Kassakian's judgment, for all other purposes "the only other energy source sufficiently dense to meet our needs, sufficiently benign for minimal ecological impact, and in sufficient quantity to free the West from the politics of the Middle East" is nuclear (uranium) fission.

"The world has said 'no' to nuclear energy," Kassakian acknowledged, "but this has to be viewed as a temporary deferral. In the twenty-first century, nuclear energy is going to *have* to be explored in a rational way." Developing nuclear power further "is a social challenge, not a technical challenge," he said. "The waste and safety issues are solvable problems. But how can we get the emotion out of public debate and get the facts into it? How can we convince people that an emotional response may not be based on *fact*?"

John G. Kassakian, who received his IEEE Fellow award in 1989 "for contributions to education and research in power electronics," was the founding president of the IEEE Power Electronics Society in 1988. He is a co-author of the textbook Principles of Power Electronics *(Reading, Mass.: Addison-Wesley, 1991). In 1997, he was elected to the Board of Directors of ISO New England, which has responsibility for the operation of New England's electric power generation and transmission system.*

at—and even shut off some cylinders when you're cruising at highway speeds." The result will be "improved engine power, efficiency, and emissions control, which will help benefit the environment." [See his box "How soon can we free ourselves from fossil fuels?" p. 34.]

The introduction of power electronics for the individual control of the pumps for power steering, water, and air conditioning means that those three devices will not all be running off one accessory belt (fan belt) at the front of the car. Thus, the "packaging of the engine in the car becomes less constrained, and gives the automotive stylist more freedom in dealing with the front of the car."

Technician Bob Louis of Keithley Instruments, Cleveland, Ohio, fine-tunes what soon may be standard under the hood: miniaturized instruments in a new in-vehicle electrical monitoring system—shrunk down from larger laboratory instruments—to take data on automotive electronics, as well as on engine and transmission temperature, engine speed, pressure, strain and vibration. Increasingly, such devices will incorporate new power electronics. *Feature News Photo and Keithley Instruments*

And the advent of electromechanical suspension, where each of the four struts is individually controlled through power electronics linked to accelerometers, will mean "you can go over a log road and not even feel it." He added: "Now, *that's* a dynamite innovation where the consumer will indeed find a visible benefit!"

Behind the Scenes II: The Millimeter Revolution

"I anticipate there will be a huge potential for millimeter and sub-millimeter applications in the twenty-first century," said the University of Leeds' Roger D. Pollard. "It's the most underutilized area of the electromagnetic spectrum."

The millimeter and sub-millimeter regions of the spectrum, at frequencies of roughly 60 to 300 GHz, is bordered at lower frequencies by the microwave region (from roughly 1 to 60 GHz), widely used for microwave ovens, cardiac pacemakers, various radars, and wireless telephones. At higher frequencies are far infrared (300 GHz to 30 THz) and infrared (30 THz and higher).

Historically, the millimeter and sub-millimeter regions' potential has been difficult to tap. "As people run out of bandwidth at radio and microwave wave-

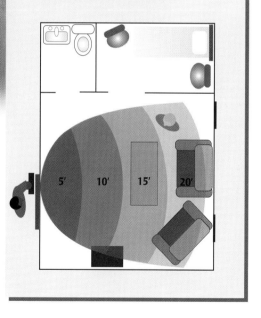

Microburst radar developed by Time Domain Corp. holds the potential for locating virtually anything, anywhere. Security applications include scanners that would check passengers and luggage alike for concealed explosives or, as shown here, tell police teams the exact locations of suspects inside a locked room.

Charles Siefried (top) and Time Domain Corp.

lengths, they've gone to higher and higher frequencies," explained Pollard. "But at millimeter and sub-millimeter frequencies, radio frequency and microwave components become so small that the problem of making them becomes really challenging. Conversely, for people using optical techniques at infrared wavelengths, going down in frequency—which corresponds to increasing wavelength—means that optical components become so large that even if you can manufacture them, the benefits of using optical technologies are lost. As a result, the millimeter and

ROGER D. POLLARD: If you could 'uninvent' a technology, which would it be?

"If you could 'uninvent' a technology, which would it be? And why?" asked Roger D. Pollard, deputy director of the Institute of Microwaves and Photonics at University of Leeds in England. Only a few moments of this intellectual exercise are needed for it to become abundantly clear just how many apparently unrelated human activities and economies either support or are supported by technology.

His own choice would be spreadsheet software. "Spreadsheet software has led practitioners in the humanities, social sciences, and liberal arts to believe that to be numerically literate, complete, and accurate, all you need to do is feed data into the rows and columns of a spreadsheet. Many decisions taken by managers who have no scientific training are based on erroneous conclusions drawn from such analysis."

sub-millimeter regions—which are neither radio nor optical—have remained underutilized," Pollard said.

But their relatively small wavelengths "have fabulous potential for building very compact systems with enormous bandwidth, making them ideal for transmission of high-resolution images," Pollard pointed out. "They also have great potential for penetration [of material objects], and do not have the ionizing problems of X-rays. There is the possibility, for example, that they can be used to detect plastic firearms that do not set off airport metal detectors—so your millimeter or sub-millimeter waves could scan people at airports, whereas you wouldn't do that

Pollard started thinking about his "uninvent" party game after "reading an entertaining science fiction novel some years ago," he recalled. "It was called *The Difference Engine* by William Gibson [San Francisco, California: Spectra Books, 1992]. It is based on the argument that electronics never got invented, and that all computers were developed from [British mathematician Charles] Babbage's early nineteenth-century mechanical difference engine. In the novel, all the things we now use exist, but are based on mechanical computers—for example, credit cards consist of bits of plastic with holes, and communications are carried through cities in connected pneumatic tubes. It was fascinating, and revealed how dependent our economy is on electronic computers."

The novel and the "uninvent" game also revealed to Pollard how the advent of a major new technology—widely hailed as creating new jobs and opportunities—often dis-

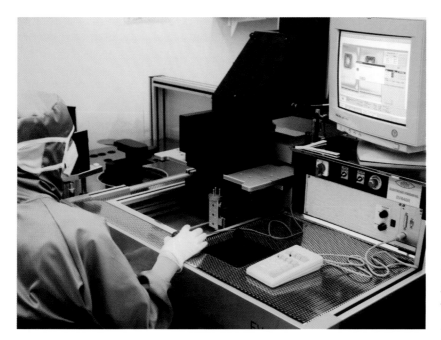

In September 1998, BREED Technologies, Inc. opened FAB 2, the world's largest facility dedicated exclusively to manufacturing silicon micromachined sensors such as airbag triggers. Yet technicians such as this one may face retraining or unemployment if the promise of nanotechnology is realized and microtech goes the way of vacuum tubes. *PR NewsFoto and Breed Technologies*

with X-rays as they are too dangerous." Their potential for imaging nonmetallic materials also make them useful for applications in medicine and remote sensing.

But one of their most promising applications is vehicular radar at 77 GHz—a radar collision-avoidance system "the size of a shoebox or smaller, that will give enough resolution to track the car in front of you and still distinguish it from cars in other lanes," Pollard said. Such systems have been feasible for at least 20 years, but only at prohibitive cost—tens of thousands of dollars. New technology, however, is making such radars possible with commercially acceptable costs in the hundreds of dollars. Such systems may be commercially available in upscale cars

'Engineers can invent almost anything you want. The question is: Are you going to make *good* use of it?'

enfranchises whole economic sectors dependent on an older technology. "Shortly after the turn of the twentieth century, someone wrote an editorial arguing that the automobile should be banned in England because of how many people who work with horses would be put out of work—saddle makers, stables, feed suppliers, carriage-builders, buggy-whip manufacturers."

In his mind, the lesson is that engineers need to be sensitive to how inventions may create ills and dislocations, as well as boons, to humanity.

"I don't work in this business solely because I find it intellectually interesting," he concluded. "I believe that technology has a lot to contribute to society. I tend to take the view that engineers can do almost anything you [society] want.

"The question is: Are you going to make *good* use of it? And is it any of my business if you won't?"

◆

Roger D. Pollard, who received his IEEE Fellow award in 1997 "for contributions to the development of microwave and millimeter-wave measurements, and active device characterization," holds the Hewlett-Packard Chair in High Frequency Measurements at the School of Electronic and Electrical Engineering at the University of Leeds. He is the author of more than 100 technical papers and a holder of three patents.

---

as early as the 2000–2001 model year, and are being designed into mid-range cars planned for 2005 to 2010. Ultimately he expects them to be in half of all cars.

Essentially, the radar will provide the driver with an "autonomous cruise control," Pollard said. "You don't have to control the distance between your car and the one in front. The radar instead controls it by time—when the car in front slows down, you do, too; when it speeds up, you do as well, increasing the separation so you always have adequate braking distance." He does worry, however, that some drivers may become lulled by its tracking effectiveness and be tempted to drive 70 miles per hour through thick fog: "The weakest part of a motor car is

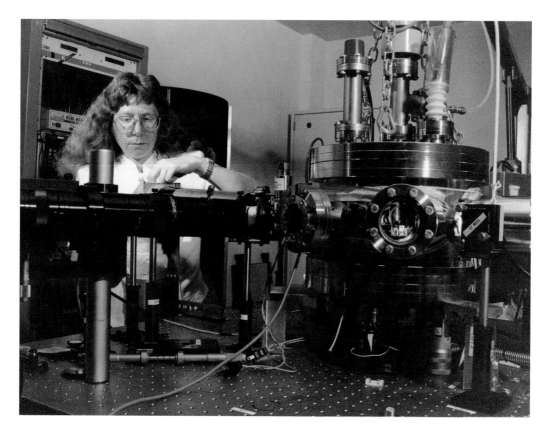

Jan Rogers at NASA's Marshall Space Flight Center prepares to test a BB-sized sample in the Electrostatic Levitator. The precise physical properties of many promising high-performance metals and ceramics are unknown because the materials react with their containers when molten at high temperatures. The ESL allows "hands off" measurements of viscosity, surface tension, and even supercooling—cooling below the normal freezing point—which can lead to the formation of glasslike metals.
NASA/Marshall (top), Loral Space Systems

the nut holding onto the steering wheel." [For Pollard's perspective on the pervasiveness of technology in society, see his box "If you could 'uninvent' a technology, which would it be?" p. 40.]

Semiconductor Diode Lasers: Industrial Tools

"Semiconductor diode lasers are to optics as silicon transistors were to electronics forty years ago," stated SDL's chairman and CEO Donald R. Scifres. Specifically, Scifres believes that the semiconductor laser can replace most, if not all, laser technologies over the next 30 years, just as the transistor—and subsequently

DONALD R. SCIFRES: How will information technology transform global culture?

"I'm not clairvoyant, but one thing is certain: communications services—the Internet, the telephone, television, and other information technologies—will force cultural changes around the world," stated Donald R. Scifres, CEO, president, and chairman of the board of SDL Inc. in San Jose, California. Scifres and his company have pioneered commercial developments in high-power semiconductor optoelectronics and laser technology for optical communications.

"The combination of computers and communications becoming less expensive and the Internet becoming more pervasive will create lessened differentiation among people around the world. Information technologies will make it possible for people to get educational degrees in every country and to keep up-to-speed on technical trends, just by getting on a computer, instead of having to live in the right country," he said.

the silicon integrated circuit—replaced vacuum tube technology of the early twentieth century.

In industrial applications today, there are several coexisting laser technologies, Scifres explained. They include high-power gas lasers (such as carbon dioxide lasers used for the cutting or welding of thick metal plates), somewhat lower-power solid-state lasers (such as neodymium yttrium aluminum garnet—abbreviated Nd:YAG—used for hole-drilling, marking of surfaces, and machining of thin metals), and low-power semiconductor lasers (used primarily for fiber optic communications, and for reading and writing on a rotating optical disk memory).

"Information is very powerful. Part of the reason the former Soviet Union dissolved is that information technology made it impossible any longer for thoughts and ideas to be controlled by political leadership. Technology allowed the ideas to spread. Even though China and other countries may attempt to control access, that will be difficult to do, because people in those nations will want more rights, just as those in the former Soviet Union did.

Who needs art museums when you can have the virtual statue in your office? IBM has developed digital modeling methods that will let art historians study fragile artworks without touching them. Any portion of the statue can be viewed from any angle and any lighting conditions that the historian selects. In the museum, the views and lighting accessible to the historian are restricted. IBM is developing new methods to capture and render additional color and texture information to increase the fidelity of these synthetic renderings. *IBM*

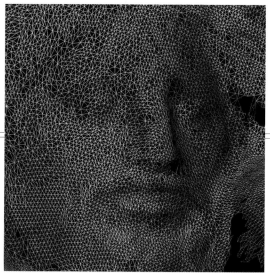

Individual semiconductor diode lasers, which are approximately the size of a grain of salt, still seem fundamentally limited in their power capability. Today the tiny diodes can emit up to only 300 milliwatts—less than a third of a watt—from a small spot 1 micrometer by 3 micrometers square. "Thus, on the face of it, they appear to be unlikely choices for the toughest cutting and welding applications, which demand kilowatt-plus power levels," Scifres said.

'It is not necessarily good for the world to have one common culture, language, and financial structure.'

"Will information technologies lead to a common global language and cultural values, with all that Western materialism implies?" Scifres asked. "That's an interesting thought—the Internet is tending to promote English as a common language because so far the Internet is largely in English.

"But it is not necessarily good for the world to have one common culture, language, and financial structure. The concerns of those who fear the loss of diversity and the death of unique cultures and values and spiritual security are well-founded. We've already seen cultural backlash against such homogenization in Canada (Quebec) and France.

"Moreover, technology does not always mean a better life for people," he cautioned. "For example, it is certain that the pollution in China resulting from increased demand for more automobiles and infrastructure technologies will be a significant problem of global proportions."

Donald R. Scifres, who received his IEEE Fellow award in 1981 "for contributions to the science and technology of diode lasers," is a former president of the IEEE Laser and Electro-Optics Society. He also is a member of the National Academy of Engineering, a Fellow of the Optical Society of America, a holder of more than 120 U.S. patents, and the author of more than 300 technical articles.

Large arrays of semiconductor diode lasers, however, can be focused to high brightness by arrays of micro-optics. And such arrays are now beginning to be operated at powers of greater than a kilowatt to achieve direct materials-processing applications similar to those achieved by the highest power lasers around. "In the late 1990s, for example, an array of semiconductor lasers generating 1.5 kilowatts welded 1.25-millimeter-thick stainless steel and surface-hardened carbon steel," he recounted. "These applications are barely achievable with Nd:YAG solid-state lasers and were unthinkable 10 years ago for diode lasers."

Semiconductor diode laser arrays are attractive for materials-processing applications for the same reasons that silicon transistors are preferred over vacuum tubes: They are much more compact than other types of lasers. They are much more reliable. And they operate at power conversion efficiencies of 5 to 50 times higher. "Up to 66 percent electrical to optical efficiency has been reported for diode lasers—a world record for any type of laser and very close to the limits imposed by fundamental physics," Scifres pointed out. (For comparison, typical carbon dioxide gas lasers have an efficiency of about 5 percent.) Moreover, semiconductor lasers can be configured to operate at wavelengths ranging from the near infrared to the visible—wavelengths that generally couple well to the materials to be heated.

One major technical challenge remains, Scifres noted. So far, the brightness of the semiconductor lasers that can be focused with micro-optics—tiny lenses configured on the laser's crystal face or on separate sheets of glass—is limited. This limitation causes short depths of field—meaning that the laser has to be very close to the work piece instead of being able to be held tens of centimeters away—to achieve a power density sufficient for all types of materials processing. However, many researchers are investigating ways of making arrays of lasers emit coherent radiation—light that is all in phase, as with an individual laser. "With high-power *coherent* optical power," Scifres said, "semiconductor diode lasers could overcome one of the last hurdles in their replacing most other types of lasers used in industry today." [For Scifres' views on the information revolution (brought about in part by low-power semiconductor lasers), see his box "How will information technology transform global culture?" p. 44.]

Molecular Manufacturing

"Manufacturing is about rearranging atoms," remarked Xerox PARC's Ralph C. Merkle. Molecular nanotechnology—literally the building up of an object atom by atom, much as one might arrange a child's LEGO® blocks to create a building or machine—"is the ultimate manufacturing technology. It gives the ultimate precision, with almost every atom in the right place. It gives the ultimate in flexibility consistent with physical law. And ultimately, it will be low cost, less than U.S. $1.00 per kilogram, roughly the price for wood."

No, such manipulation of individual atoms does not violate physical law. In fact, it was even demonstrated in a crude proof-of-concept experiment in 1990 when IBM scientists used the tip of a scanning probe microscope to arrange thirty-five xenon atoms on a nickel surface to form the three letters of their employer's name.

The ability to arrange some atoms with a scanning probe microscope does not prove that twenty-first-century manufacturers will be able to construct com-

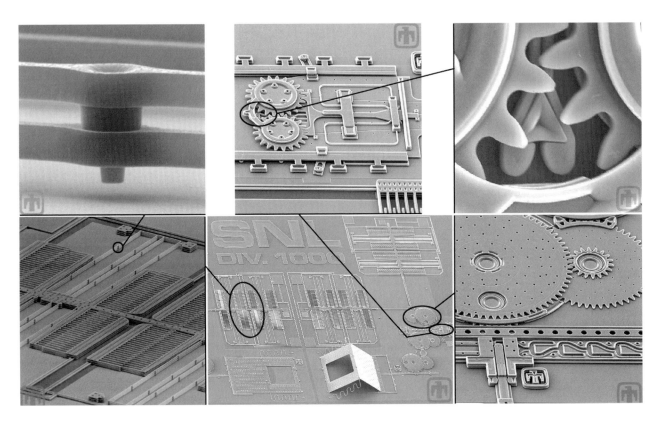

Leading the way for nanomachines will be a range of micromachines that will be manufactured using technologies derived from electronic microchip production. Engineers in several laboratories are building devices that may seem to serve no useful purpose (yet), but develop a new range of mechanical engineering skills. This device, shown under a scanning electron microscope, comprises a complex locking mechanism that eventually raises a small mirror on a hinge. *Sandia National Laboratories, Intelligent Machines Initiative (*http://www.mdl.sandia.gov/Micromachines*)*

plex structures and machines by placing atoms and molecules one at a time. But biological systems, which do it as a matter of course, demonstrate that remarkably complex structures are not only possible, but can be economical as well.

Merkle points to the example of trees. "Trees grow by taking energy from sunlight and nutrients from the soil to build themselves. They do it quietly and discretely, without digging holes in the ground or burning fossil fuels or throwing off noxious waste products. They use only what they need, arranging the atoms in complex internal patterns. And trees also self-replicate: They produce seeds that build other trees. Precisely because it's a miracle of biology, lumber costs only a few dollars a pound."

Living systems use a powerful method for arranging atoms: self-assembly. A basic principle in self-assembly is what Merkle calls "selective stickiness." If two molecular parts have complementary shapes and charge patterns, then they will tend to stick together in a particular way to form a bigger part. Selective stickiness could also be useful for building early "nanotools" that would then be used for building other things. Merkle and others propose molecular-scale positional devices for holding molecules in precise position, and one-ten-millionth-scale robotic arms that might sweep back and forth over a surface, systematically adding and withdrawing atoms to build any structure.

Self-replication is also essential, because "once you have a technology that can make copies of itself by placing individual molecular parts where they're needed, then you also have a manufacturing process that is intrinsically low cost." For example, if you want to make a relatively large object, such as a car or a supporting beam for a building, the time-efficient approach, Merkle noted, is to build it in steps: First, use nanotools to build an assembler—a small, computer-controlled robot that can be programmed to build almost anything. Then program the assemblers to replicate themselves. Last, when you have enough assemblers, build the product.

"As the binding times of molecules is brief, the speed of manufacturing is limited by the speed that the tools can move molecules into position. If a tool could move a molecule into position in about a microsecond, then you would have a million operations per second—and since an assembler is composed of about a billion atoms, it would take about 20 minutes for an assembler to build a copy of itself," Merkle calculated. But then the power of geometrical progression—doubling every 20 minutes—would let you make a billion billion assemblers in just 60 replication periods, or 20 hours.

Then all the assemblers would work simultaneously on assembling a product. "Many hands make light work," Merkle observed. "You could build a car in a few hours"—comparable to the time required by traditional manufacturing processes.

To keep things relatively simple, Merkle has been focusing his theoretical investigations specifically on the properties of hydrocarbons—molecules composed of atoms of carbon and hydrogen. Not only do some hydrocarbons have highly useful structural and electronic properties, but carbon compounds are also compatible with human biology (for more about the potential of molecular nanotechnology in medicine, see Chapter 7: Medicine and Biology). Also, he noted, "diamond and its shatterproof variants fall within this category, as do 'fullerenes'—carbon atoms that naturally form into spheres, tubes, and other shapes needed for basic mechanical devices such as struts, bearings, gears, and robotic arms."

How long will it take for nanotechnology to become practical reality? "The scientifically correct answer is: I don't know," Merkle said. "Ten years? Fifty years? I think it will be faster than that. There is no reason in principle we should not

RALPH C. MERKLE: Are we prepared for the nanotechnology revolution?

RALPH MERKLE

"Molecular nanotechnology will have a tremendous impact on all aspects of life in the twenty-first century," declared Ralph C. Merkle, a research scientist at Xerox Palo Alto Research Center (PARC) in California. Merkle, one of the field's pioneers, was named the 1998 co-recipient (along with Stephen Walch of the Thermosciences Institute at the U.S. National Aeronautics and Space Administration's Ames Research Center in Moffett Field, California) of the Foresight Institute's 1998 Feynman Prize in Nanotechnology for Theoretical Work.

According to Merkle, among the benefits of mastering molecular nanotechnology—a proposed process for building large structures by the precise manipulation of individual atoms and molecules—will be not only major reductions in manufacturing costs, but

have at least some practical molecular manufacturing capabilities by 2010 or 2020. It all depends on how soon we start!" [For Merkle's assessment of what molecular manufacturing would mean to society, see his box "Are we prepared for the nanotechnology revolution?" above.]

Predictions: Economical Colonization of Space

"One application of molecular nanotechnology would be reducing the weight of rockets for boosting payloads into orbit," Merkle said. "The strength-to-weight ratio of diamond is about fifty times greater than that of steel or aluminum alloy.

also reduction in twenty-first-century society's dependence on fossil fuels, reduction in environmental pollution of all kinds, and increased food production for the world.

With molecular nanotechnology, the manufacturing process could come close to its theoretical maximum efficiency and minimum waste. Thinking of the scrap and shavings produced by cutting metal or wood, Merkle observed: "There is nothing fundamental about spewing out random undesired atoms to the environment as part of the manufacturing process."

Since molecularly-tailored materials similar to diamond could well have internal structures with strength-to-weight ratios 50 times that of steel or aluminum alloys, products such as cars may have characteristics very different from those today—"you could have a

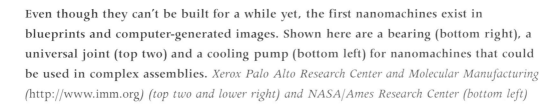

Even though they can't be built for a while yet, the first nanomachines exist in blueprints and computer-generated images. Shown here are a bearing (bottom right), a universal joint (top two) and a cooling pump (bottom left) for nanomachines that could be used in complex assemblies. *Xerox Palo Alto Research Center and Molecular Manufacturing (*http://www.imm.org*) (top two and lower right) and NASA/Ames Research Center (bottom left)*

'Nanotechnology might bring over a factor of 10 improvement in the yield per acre in agriculture.'

100-pound Cadillac, or a $1/2$-pound bicycle," he speculated.

The economics of manufacturing physical objects would be completely revolutionized, coming more to resemble the economics of producing software—where the cost of stamping a CD-ROM is very low compared to the cost of developing the content. "Manufacturing costs will shrink to the minimum required for making the product itself" rather than including the cost of all the waste, electric power, manpower and overhead, he observed.

Another implication is that massive central smokestack factories may become a relic of the past. Since products would be constructed by microscopic self-replicating assemblers, and since the process would be so clean, consumer products could be manufactured in neighborhood centers, minimizing the need for products to be trucked nationwide through distribution channels. In fact, Merkle thinks that small household products such as toasters could "maybe even be manufactured in individual households." A person might buy or rent a manufacturing machine, fill it with a feed stock that Merkle envisions might be rather like toner for a copying machine, program the machine to produce a toaster, and let it run overnight—and next morning, a toaster is ready-made for use.

Such low-cost manufacturing could also make inexpensive solar cells and batteries, letting cheap and plentiful solar power replace coal, oil, and nuclear power, and "minimizing the need to dig holes in the ground" for resources, he said. "And because materials similar to diamond could literally be made from the carbon in the air—just as trees get their carbon from carbon dioxide—there would be little need to strip-mine for metal."

If you could reduce the weight of a rocket by a factor of 50 by using shatter-resistant variants of diamond, and if the manufacturing costs of that material were about $1.00 per kilogram, then you should be able to reduce the cost of putting things into space by three to four orders of magnitude."

The U.S. National Aeronautics and Space Administration (NASA), realizing its potential, has a nanotechnology research group at its Ames Research Center in Moffett Field, California, exploring (among other things) the possibility of "using indigenous materials to live off the land on the Moon or Mars," Merkle pointed out. "The idea is to allow scientists to build everything they would need for a

Molecular nanotechnology might even bring about "over a factor of 10 improvement in the yield per acre in agriculture, allowing you to reduce the amount of acreage to grow food," he suggested. "If molecular manufacturing allows the construction of low-cost greenhouses, farmers could precisely control the environment in which plants grow. They could also optimize those conditions for growing year-round." Thus, crops would not be subject to the vagaries of dry years, pests, and other causes of failures that plague crops grown outdoors.

'Costs of manufacturing physical objects will shrink to resemble the economics of producing software.'

"There will be substantial shifts in economic sectors," Merkle hypothesized. "Companies that primarily manufacture products might find themselves in a difficult position, while those that specialize in design might find themselves in a good position."

◆

Ralph C. Merkle was a co-inventor of public key cryptography in the early 1970s, for which he and five other researchers received the Association for Computing Machinery's 1996 Paris Kanellakis Theory and Practice Award. He is an editor of the Institute of Physics Publishing's journal Nanotechnology, *the author of more than 50 technical papers, and the holder of eight U.S. patents.*

permanent colony from what's available, rather than trying to lift everything needed off the earth.

"Moreover, at such low cost, you and I could afford to go into space—not just to take a vacation, but to explore a new frontier," Merkle added. For that very reason, he pointed out, the National Space Society—a decades-old space interest group based in Washington, D.C.—"has many people interested in molecular nanotechnology as a means for advancing the human use of space. They see it as an opportunity for them, personally, to participate in the exploration of space rather than just to watch a handful of astronauts do it." ◇

CHAPTER 3

SYSTEMS AND MANAGEMENT

THE LIVES OF INDIVIDUALS AND THE LIVELIHOOD of society are increasingly dependent on electronic devices and systems. Thus, the quality and reliability of electrical power is the frontier for power utilities in the twenty-first century and is more crucial than ever before, stated Robert A. Bell, recently retired vice president of research for the Consolidated Edison Company of New York. Samuel J. Keene, chief technical officer of Performance Technology Consultancy in Boulder, Colorado, added that reliability is also increasingly important for enabling new systems—especially those that affect human life and safety.

To appreciate these concerns, engineers must take more of a systems view and must become more skilled in managing technology from conception to disposal, said Wade H. Shaw, Jr., professor of engineering and technology management at the Florida Institute of Technology in Melbourne, Florida. In fact, he contends, life-cycle management should be viewed as integral to the technology.

Eventually, the day may come when the hierarchical network structure of large-

Left: If managers are to re-tailor the shape of industries and transportation to avoid global warming, then it helps to understand exactly what warming will do in addition to raising sea level. In the middle of a sweet gum stand along the Clinch River in Tennessee, the Oak Ridge National Laboratory has built aluminum towers for the Free Air CO_2 Enrichment experiment to determine how forests and ecosystems react to increased carbon dioxide in the environment. The elevated carbon dioxide levels are within ranges beneficial to plants. *Oak Ridge National Laboratory photo by Steve Eberhardt*

scale distributed information systems may actually augment the intelligence of just plain folks, hypothesized Rui J. P. de Figueiredo, director of the Laboratory for Machine Intelligence and Neural and Soft Computing at the University of California at Irvine.

Power System Evolution

"In the energy business, we're still bound by the thermodynamic laws of Carnot and Rankine," said Consolidated Edison's Robert A. Bell. Carnot's law states that the thermal efficiency is determined by the difference between the highest and lowest temperatures in the cycle, and Rankine's law takes into account the steam conditions that are utilized. Both laws drive designers to use the highest temperatures or pressures that materials will tolerate over the life a power plant. "We still need to combust fuels to heat air or raise steam to drive turbines and generate electricity. Our [technological] progress is bound by materials limitations and thermodynamic laws, and is thus incremental in nature. So we don't anticipate a technology revolution like that in computers and communications."

Nonetheless, Bell feels "there is still a lot of progress to be made."

In his opinion, deregulation of the power industry and the introduction of competition—spearheaded first in the United Kingdom, next in the United States, and then spreading to Germany, Japan, and Sweden—"will lead to greater efficiencies and quality of service." As power generation will no longer be regulated, there will be increased pressure on plants using various fuels (natural gas, coal, nuclear fission, hydroelectric) to reduce cost and increase efficiency. "The plants that survive will be the most economically efficient ones that provide the lowest-cost energy to consumers," Bell predicted. [For Bell's thoughts on the environmental costs of various fuels, see his box "When will society recognize that nuclear reactors are environmentally safer than fossil-fuel power plants?" p. 58.]

One development that will increase efficiency is the active management of transmission and distribution of electric power, he said. For the last century, transmission and distribution circuits have relied on passive electrical components. At the generator, voltage was maximized and current minimized to reduce transmission losses on the lines, while at the destination, the voltage was stepped down to a practical level before entering businesses and homes. "The most active components were circuit breakers or switches," Bell recounted. In contrast, communications networks have amplifiers and extensive signal-processing electronics for optimizing transmission of information and data.

With advances in power electronics [see Chapter 2: Structures and Devices], the development of switching devices that can tolerate the voltage and current

Airbus Industrie's A319 and A321 jetliners undergo their final assembly at the Daimler-Benz Aerospace Airbus factory in Hamburg, Germany. The aircraft are assembled from parts built across the globe that must be brought together with exacting tolerances. The flow of parts must also be managed to ensure that they arrive when needed. *Airbus Industrie Inc.*

levels of power transmission lines has been made possible. These active techniques, known in the power industry as flexible A-C technology systems (FACTS), "are allowing utilities to load the transmission lines closer to their thermal limits," that is, the maximum temperature limits allowed by good practice, Bell said. "All that will mean is that we can better utilize existing structures and won't have to build as many new transmission lines" to accommodate growing demand from increasing population and uses, Bell explained.

Active power management will also "contribute to the reliability and security of the system," Bell added, better controlling voltage transients (spikes or drops), and minimizing recurrences of the much-publicized power blackouts in the Northeast and the Pacific Northwest in the last third of the twentieth century. With increasing reliance on computer and electronic systems of all types in business, in the twenty-first century, "society is going to place higher demands on power reliability and quality, and our tolerance for interruptions will continue to drop," he observed. New York City now could not tolerate the 12- to 24-hour blackouts it suffered several decades ago "because the financial industry and the business community at large are so heavily computerized and dependent on high capacity

ROBERT A. BELL: When will society recognize that nuclear reactors are environmentally safer than fossil-fuel power plants?

PHOTO APPEARS COURTESY OF SUSAN SMYTH BELL

"When the environmental costs of fossil fuel combustion start rising, that's when nuclear [uranium fission] power plants will come into their own again," predicted Robert A. Bell, recently retired vice president of research for the Consolidated Edison Co. of New York, in New York City. Bell's concern stems from the fact that as developing nations in the twenty-first century seek living standards that approach those of the developed Western nation, "the first thing people demand is electricity," he said. "And at the beginning, most of that electricity will be produced by burning coal."

communications," he said—to the point where many companies independently back up their key systems with stand-alone emergency generators.

Moreover, he sees that suburban and rural areas will also want the high level of power reliability already enjoyed by major metropolitan areas. "Cities are served by power networks or grids that are fed from multiple points," he explained. Thus, a break in one part of the grid seldom interrupts power to big areas because there are alternate sources and connecting paths. Also, power lines in cities run underground, where they are not subject to lightning strikes, ice storms, or automobiles running into poles, as are the less expensive overhead power lines common in more

Of the fossil fuels, natural gas—methane (CH_4)—is the cleanest, with its combustion byproducts being primarily water and only a small amount of carbon dioxide, Bell explained. The carbon content in oil is much higher, so oil-fired power plants produce significantly more carbon dioxide, the "greenhouse gas" implicated as the chief culprit in contributing to the potential for global warming. But coal, "which is almost pure carbon," is the worst of all, producing carbon dioxide "as its primary waste product," Bell explained. "But since it's so cheap and widely available compared to natural gas in developing countries, it will likely be the fuel of choice."

'The key issue is the general *perception* of risk rather than the level of actual risk.'

At some point, however, when the amount of carbon dioxide in the atmosphere builds up to the point that it is pressuring climate change, "nuclear will start evolving again," he said. "From an engineer's viewpoint, nuclear represents a good, solid option, and all of its dangers readily yield to engineering solutions."

In Bell's opinion, "the key issue is the general public's *perception* of risk rather than the level of actual physical risk." Many of the fears of the general public about nuclear reactors are based on inadequate understanding about the effectiveness of good engineering design, as illustrated by two widely publicized nuclear power plant accidents in the 1980s.

At the Three Mile Island nuclear plant located in eastern Pennsylvania, an operational error caused a massive meltdown of the entire reactor core—yet the design of the reactor's safety systems was so effective that "the consequences of the runaway reaction were so well contained that no significant radioactive releases escaped into the atmos-

sparsely populated areas. Commonly, the power systems outside of cities are only simple loops or radial branches, with power fed in from just one location—so a break in one part of the system can affect many customers.

"What I see over the next 100 years is a march toward underground systems for both transmission and distribution, and the development of grid networks" outside of cities to gain greater reliability, Bell said.

Systems Reliability

"As engineering systems—such as information technology—become more

phere," Bell recounted. Although public fears of a runaway nuclear meltdown breaching all of the levels of containment were then high as a result of the plot premise of the blockbuster motion picture *The China Syndrome*, released three months earlier, "not one person was injured due to unintended radiation exposure," Bell said.

"Contrast that with the absolute disaster of Chernobyl in the Ukraine," he continued. Chernobyl was "a graphite core reactor that had *no* containment whatsoever," because it was believed that the chances of a runaway condition occurring while operating within normal ranges were so remote that the construction cost of containment was not justified. The Chernobyl accident was triggered by electrical engineers who were conducting an experiment to determine to what extent the inertial stored energy of the rotating generator could be utilized to power plant auxiliary equipment if primary electric power had been interrupted, Bell explained. Inadvertently, their experiment led the reactor to operate in ranges not only outside the design criteria, but also in an inherently unstable region. Compounding the problem was the reactor's design: the cooling-water pumps were driven by electricity instead of steam, as is the U.S. practice. As the engi-

For most engineers, delta-t means the change in temperature. Here it means the change in time as computer systems approach the year 2000 (Y2K) and the possibility that legacy programs may falter when their dates roll over to 2000—which they will read as "00." Akcess2K's Delta-T Probe is designed to test Y2K compliance in older embedded systems in the field or on a test bench, and identify chips that have Y2K risks.
PRNewsFoto and Akcess2K Inc.

neering experiment curtailed the electricity available to the pumps, the cooling water was reduced to the reactor—which was already entering into an unstable operational regime, Bell continued. And the reactor core exploded. Radiation releases were worsened when emergency crews tried to control the resulting fires by pouring water on them, creating large clouds of radioactive vapor. Without containment, highly radioactive debris and water vapor shot upward into the atmosphere, where high-level winds carried them long distances—leading to the contamination of vast areas of land.

> 'From an engineer's viewpoint, nuclear represents a good, solid option, and all of its dangers readily yield to engineering solutions.'

"If you compare the cases of Three Mile Island with Chernobyl," said Bell, "the two outcomes were dramatically different" and illustrate the consequences of safety observed versus compromised. Unfortunately, that "crucial distinction is lost on much of the general public, leading to an unfounded fear of properly designed and operated nuclear plants," he remarked.

Second-generation reactors being designed today are even more secure than the design of Three Mile Island, which is widely used in other currently-operating nuclear power plants, he said. Among other things, for cooling, second-generation nuclear power plants use a passive convection system that does not depend on the operation of pumps; thus, it can provide long-term emergency cooling for the core even in the event of a complete failure of plant auxiliary systems. "Second generation nuclear plants will carry an already extremely safe and reliable technology to even lower levels of risk—even approaching zero," Bell stated.

pervasively intertwined with our lives and enterprises, reliability is going to become increasingly important," stated Samuel J. Keene of Performance Technology Consultancy. As the first global example, he cited the billions of dollars spent worldwide by industries ranging from electric utilities to financial services to ensure that their information technology infrastructures were "Y2K compatible"—that is, would not shut down or behave unpredictably when the year 1999 ticked over to the year 2000. The much-vaunted Y2K problem, he contended, shows the overriding impact of software on today's systems.

While hardware "typically is based upon parts that are readily available and

Regarding the disposal of waste, it is not generally appreciated that "there's a significant waste disposal problem associated with coal," said Bell. "Coal combustion leaves copious quantities of ash that is contaminated by mercury and other environmentally hazardous elements." Although the radiation and long half-lives of fission products require nuclear waste to be processed into a form insoluble to water, stored in casks capable of maintaining their integrity for centuries, and placed in stable geological structures, "nuclear volumes are very small compared to waste from coal," Bell said. Moreover, it is "well documented in crash tests" that the nuclear casks are "held to an extraordinarily high standard compared to containers for commodities like ammonia or oil" in the event of a train derailment or other catastrophe in their transportation.

"When you factor in the environmental impact and the human health hazards of the strip-mining process, there's no question that the long-term risks of coal-burning outweigh the risks of nuclear-generated power," Bell concluded. "I think that as time goes on, society will balance out the environmental consequences of the uncontrolled burning of immense amounts of coal versus the consequences of well-designed nuclear power plants, and will realize that nuclear power *has* to be a key element worldwide."

Robert A. Bell, who received his IEEE Fellow award in 1992 "for leadership in the formation and management of the research and development organizations of the electric power industry," founded the Research Department at the Consolidated Edison Company of New York, an activity he led for more than 28 years. A resident of New York City, Bell holds the Electric Power Research Institute's Lifetime Achievement award (1998) and many other awards, as well as the U.S. Army's Legion of Merit.

tested, software is still mainly a custom art form," Keene observed, and is the "chief reason for program slips, cost overruns, and development program cancellations," he said. "Software-run systems are becoming more complex and entrusted with our health, safety, and well-being—including medical equipment, transportation systems (plane, train, and automobile), nuclear power control, environmental impact and enterprise systems. The users—you and I—want to know 'Can we really rely on the system control or solution?' 'Can we verify its truth?' 'Can we be assured of its robustness so that it will still function well in its operating environment despite variations in use, and will not fail catastrophically if there are faults?'" Keene said.

Ultimately, any technical organization is only as good as the teams that compose them. Here, junior executives from Samsung, the South Korean electronics and manufacturing giant, go through radical training exercises at a specially designed human resources wilderness training center in Kyonggi-Do. The four-week training program exaggerates conventional workplace challenges and encourages creative solutions to typical office-related problems. All of Samsung's executives and managers are required to pass through this program, which includes physical training, theatrical role-playing and team-building activities. The program now graduates more than 5,000 executives each year. *Feature Photo Service and Samsung*

"Software reliability is heavily contingent on the quality of the development *process* assuring thoroughness of thought and consideration," he went on. "Planned incremental change release cycles are generally a good and reliable way to grow a complex system. Problems occur when too much function change is attempted in one release. Incremental releases allow the user to gain experience with the system and to feed back requirements changes in a systematic manner." That being said, he cautioned that small changes made in response to problems "tend to be the most fault-prone, simply because code maintainers often don't give sufficient thought to small changes."

Concern about trustworthy software and systems led to the establishment of

SAMUEL J. KEENE: How can effective communication help engineers develop the best products?

"The biggest problem in product development is communication," declared Samuel J. Keene, chief technical officer of Performance Technology Consultancy in Boulder, Colorado, a firm that specializes in the newest reliability engineering practices. "Engineers are often not the strongest communicators. As a result, they too often don't talk with their counterparts in materials and manufacturing."

In Keene's experience, the best products have been built by cross-functional teams that were given tools to maximize communications. Specifically, experts from different disciplines were gathered together in one common location and were focused on one goal: designing a product to be as simple, reliable, efficient, and cost-effective as possible.

"Since good communication is the seed of the problem, it's also the seed of the solution," said Keene, who has long consulted and conducted seminars on team-building

the Software Engineering Institute (SEI) at Carnegie Mellon University in Pittsburgh, Pennsylvania, in 1984. SEI is a research and development center sponsored by the U.S. Department of Defense. SEI has developed a five-point rating scale for evaluating and promoting the best software development practices—as the better the development practice used, the better the ensuing software reliability is. Moreover, the Reliability Society and the Computer Society of the Institute of Electrical and Electronics Engineers (IEEE) are teaming up to establish better standards and practices for developing software for safety-critical systems. Their initiative includes certifying software developers and development processes as well as final software

and facilitation in product development. In his opinion, all parties need to share "crisp, clear understanding" on four main issues: requirements, interfaces, contingencies, and oversights.

"The biggest opportunity is in defining requirements," Keene asserted. "The U.S. Army, in fact, has reported that up to 99 percent of the product development field problems stem from shortfalls in fathoming and understanding the requirements." He has found that "all too often, engineers tend to develop products that aren't what the customer wants— *even though it's built to the customer's specifications!* The fact is, often customers don't know thoroughly what they want, even though they put out a spec. Often they need help to complete the concept and to discover the exceptions the product needs to handle."

> **'We tend to relive history unless we formalize the lessons learned from past experiences.'**

Keene's second big opportunity for improvement in product development is assuring good subsystem interfaces—which is, he noted, actually another tier of requirements having to do with the interaction of hardware and software, of both of these with testing, and of the whole product with the human user. "Since the best products are designed to accommodate the users' needs, the system should be developed to fit the human capabilities, not the other way around," Keene observed.

The third big opportunity needing good communication in product development is in handling operational contingencies. "You want a product to fail into safe modes," he said. "That doesn't happen automatically. Engineers need to think along negative paths to see how a product might fail, and envision how to mitigate the effect of failures." One powerful tool Keene recommends is FMEA: failure modes and effects analysis—"asking 'what if' questions to see how a product reacts to abnormal operating conditions," he said. "You don't want the product to do something no one ever thought about if subject-

products. A further goal, Keene noted, is to transfer best practices into academia for the systematic improvement of education in software development.

Regarding hardware, "a big part of diagnosing special problems is recognizing *patterns* between good equipment behavior and bad," he said. Thus, equipment developers will increasingly build diagnostic instrumentation into their systems. "Chips are cheap enough now that it's affordable to put in administrative logic for tracking health, behavior, and symptoms" like the black box or flight recorder in an aircraft, he said. "That way you can detect the sequences of operations that led to failure."

ed to unanticipated variations in temperature, humidity, or inadequate training of the human user."

The last major area in product development Keene has identified for improvement in communications is the catching of oversights. "We tend to relive history unless we formalize the lessons learned from past experiences, so we can apply them to the next projects," Keene said. "That's how engineers end up with silly but expensive design flaws, such as that of the Porsche model in the 1960s in which to change the sparkplugs you had to pull the engine," he said. "Throughout the design and test process, as you identify defects, you need to study them as a team, fix them, and look for similar ones elsewhere in the product. Most important, people need to learn from the experience so they don't repeat the same errors in the future."

Since all four areas of opportunity depend on effective communication, "that's why it is important—especially for a large team of several hundred people—to have team-building and project facilitation exercises," Keene said. "People need to come to know one another, feel free to talk openly with one another, brainstorm together, and develop the confidence to trust one another."

Samuel J. Keene, who received his IEEE Fellow award in 1995 "for the advancement of reliability technology, components, and systems," has produced eight video tutorials on different aspects of development, reliability, and concurrent engineering, and is the author or more than 100 technical papers and book chapters. He currently consults on reliability and safety aspects of a new flight control system for the U.S. Federal Aviation Administration, and is supporting advanced reliability technology through Six Sigma Practices for Seagate Technology.

Diagnostic instrumentation is especially valuable in ferreting out frustrating intermittent problems that are so elusive during failure analysis. "Typically, half of failure investigations of hardware report NDF (no defect found) or CND (could not duplicate)," Keene said. But a reasonable percentage of NDF failures may arise because people are using the system in a manner different from that anticipated by its designers—something that could be readily detected and recorded by diagnostic instrumentation.

"Most reliability problems are due to human oversight and error," Keene noted. That being true, a simple but effective technique is treating a detected hardware

Operating complex systems eventually comes down to faith in the other people, especially when it's your body riding on 1,750 metric tons of high-energy chemicals. A

Space Shuttle pilot (top) is introduced to the new Multifunction Electronic Display Subsystem (MEDS) that gives the Space Shuttle 1990s-level computer displays driven by older 80386 chips. While he focuses on that, he does not want to worry about the Shuttle's heatshield tiles that are inspected and sometimes replaced after each mission (below). *United Space Alliance*

failure not as a punishable offense, but "as a valuable piece of data—and then examining the rest of the system to see if a similar root cause could exist elsewhere. "Thus, the discovery of one fault can lead to the cleaning up of several other faults as well," he pointed out.

Last, in Keene's experience, engineers and managers "need to promote fuller thought—beta testing, post-mortems, data bases of lessons learned, standardized design guides, cross-functional teams where people with different areas of expertise ask 'Why?' and "Why not?'" [For more of Keene's views on the value of cross-functional teams, see his box "How can effective communication help engineers develop the best products?" p. 64.]

Management *is* a Technology

"As companies transition from traditional functional organizations—where people are grouped hierarchically by departmental function with multiple levels of management—to more flattened project-management organizations centered around cross-functional design teams, an engineer's natural career path will be from project engineer to team leader to project manager and beyond," observed Florida Institute of Technology's Wade H. Shaw. "And increasingly, team leaders and project managers are managing technology integration not just for the duration of product development, but over the product's entire life cycle, from conception through fielding and eventually disposal."

Because the management of technology integration blends the skills of traditional engineering with traditional management, in Shaw's view "management and engineering need to reinvent themselves" to be effective in the twenty-first century. Specifically, engineering "needs to become more systems oriented" and manage-

Injecting humor during difficult situations can help employees move through transitions. The Monster Board, which operates a popular careers web site (www.monster.com), installed unique "business-social" workspaces at its new office space in Maynard, Massachusetts. They include bright colors, life-size 3-D monsters, teaming spaces, a game room and an on-site workout facility. *Feature Photo Service and The Monster Board*

No matter how well motivated a production team may be, errors inevitably creep into products. Here, Theresa Long of Keithley Instruments in Cleveland, Ohio, inspects a circuit board used to test final product quality and to monitor the manufacture of PCs, wireless devices, digital consumer products and automotive electronics. These boards find product flaws by testing electronic products and the components inside, searching for extremely low electrical signals at extremely fast speeds, gathering as many as 1 million samples per second at levels as low as 0.001 volt. Demand for these systems is increasing around the world, as worldwide electronic equipment production is forecast to grow 8 percent a year through 2001. *Feature Photo Service and Keithley Instruments*

ment "needs to become less of a philosophy of human and organizational behavior and more of a practical technology itself."

In the context of engineering projects, "management skills are *part* of the technology that delivers the products and services," Shaw said. Even a comparatively simple product such as a CD player integrates vastly different technologies—power, optics, software, acoustics, mechanics. Complex technologies, such as those for designing and building a spacecraft, a new passenger airliner, or a wireless communications network, "require knowing how to integrate people, techniques, and materials in extraordinarily complex business environments, and knowing how to motivate effective teamwork and collaboration," Shaw observed. "While engineering management may include traditional management tasks, such as recruiting and designing incentive systems for employees, it also requires specialized skills

WADE H. SHAW, Jr.: Can engineers abdicate leadership forever?

"Engineers as professionals have to be willing to put themselves into leadership roles, which historically they have not done," stated Wade H. Shaw, Jr., professor of engineering and technology management at the Florida Institute of Technology in Melbourne, Florida. "Yes, some companies have been proactive in implementing a dual career ladder," he acknowledged, referring to the system of allowing parallel tracks for advancement, one along the traditional management task and the other along a primarily technical track. "But I don't see how engineers can avoid managerial responsibilities forever."

Even on the technical track, he said, engineers "still have to be concerned about cost, staff, time to market, mentoring, recruiting, and overseeing subordinates." More importantly, "courts now debate intellectual property issues that require very technical evidence. And companies make product and process decisions that must rely on engineering judgment as much as they *rely* on tax and legal advice."

Thus, Shaw urged, engineers should "seek out opportunities to exert leadership—view the opportunities as *part* of a successful engineering career."

In his opinion, if engineers willingly step into management—that is, they are willing to make decisions and take responsibility for them—they have everything to gain and nothing to lose.

---

in value engineering, cost accounting, design engineering, quality management, and project management of far-flung contractors in different time zones."

Engineering management also requires skill and comfort in making autonomous decisions, because in a project-based corporation "the team is literally a virtual organization working on a common problem until it is solved." [For more of Shaw's views on the necessity for engineers to become comfortable as decision-makers, see his box "Can engineers abdicate leadership forever?" above.]

There's just one major problem. "Little in traditional formal engineering education prepares engineers for this shift in their careers to more management re-

First, engineers will improve their likelihood for job security by "enhancing their attractiveness to their companies over the long term," he pointed out. "If your basis of influence is solely your technical skills and they're not current, then what's going to happen?" he asked. "In an era of rapidly changing technologies, when software engineers have a half-life of only three to five years, engineers will have a better opportunity for lifelong careers if they embrace leadership skills."

Second, they will be well positioned to explore broader careers in other areas that need a combination of analytical, technical, and management skills. "Engineers will find themselves popular with marketing teams, financial planning teams, executive board teams, as hospital directors, and as financial analysts for technology mutual funds," said Shaw. "They could even run for elected office."

Last, a side benefit Shaw foresees is that "to the extent that engineers are willing to integrate themselves into management teams and the organization's broader structure, the image of the engineering profession will improve to more that of the status of a medical doctor or an attorney," he suggested. Like physicians and lawyers, engineers will be valued in management because "they bring to the table a skill and perspective of value."

In short, engineers should no longer be satisfied "with just making recommendations to management," Shaw advised. "Be *part* of the solution. And be prepared to take responsibility for your decisions."

◆

Wade H. Shaw, Jr., a registered professional engineer in Florida, Ohio, and South Carolina, is chair of the Florida Institute of Technology's Engineering Management Program. He serves as the executive vice-president of the IEEE Engineering Management Society, and is a 1996–99 IEEE Distinguished Lecturer.

sponsibilities," Shaw pointed out. "Up to now, the tendency in companies has been to grab the best design engineer and thrust him or her into a management role. But there is no evidence that the best design engineer always makes the best manager. And most companies are not very tolerant of failure."

Some companies do provide on-the-job management training for promoted engineers. "But do we really think that someone can become an effective program manager by taking a weekend seminar?" Shaw asked. To address the lack of preparation of engineers for their changing roles within companies in the twenty-first century, some 120 graduate programs in the United States now offer master's de-

A NASA engineer takes her hands through the motions to stack a set of blocks with a robot. Increasingly, industry will rely on teleoperated (remotely operated) and robotic systems to perform tasks. This will impose new responsibilities on management to ensure that the automated servants do what their masters intended, which is not always the same as what the servants were unintentionally told to do. *NASA/Marshall Space Flight Center*

grees in engineering and technology management (compared to about 700 offering the traditional master's in business administration, or M.B.A.).

The important message is that "there are tools and techniques to help engineers in their new management roles," Shaw said. "To be sure, management is not a science—but neither is engineering! Managers have to take risks, experiment, put something together, try it, test it, try it again—just like engineers do. But there is a growing body of established best practices that can help."

Prediction: Collective Intelligence Through Networking

"Some of the limitations of the human brain will be overcome by the information systems of tomorrow," remarked Rui J.P. de Figueiredo of the University of California at Irvine. Based on his work, he feels that "the best possible results both individually and societally" can be obtained by people of average intelligence who augment their capabilities through intelligence distributed through a hierarchical network. In fact, he declared, "hierarchical-network-based information systems of tomorrow are likely to constitute the most significant development in the next century, if not for the next millennium."

Any network is "a set of nodes connected by links," de Figueiredo explained. A node is a processor of information. Conceptually, the node transforms a set of input signals into a set of output signals while carrying out a specific subtask. A link is a channel that transports appropriately coded signals from one node to another, modifying the signals according to the channel's properties.

A hierarchical network is a network of networks, where some or all of the nodes are themselves networks—either of other networks, or of conventional nodes (processors).

To be sure, he said, the Internet has the "anatomy" or physical structure of a global hierarchical network, whose nodes are themselves regional wide area net-

These two figures depict an analogy between the human cerebral cortex and the World Wide Web (WWW). The intelligence-oriented functions of the cortex are enabled by the hierarchical cortical neuronal network, which we call BRAINternet in analogy with the WWW's Internet network. BRAINternet is partitioned into 52 Brodmann regions. Each of these regions may be viewed as a wide area network (WAN) of the BRAINternet which, alone or in cooperation with some other regions, is responsible for some intelligence-oriented function. Thus, region 4 is the motor cortex, region 17 is the primary visual cortex, and regions 41 and 42 constitute the primary auditory cortex. To complete this analogy, the bottom figure shows the Internet network as a hierarchical network supporting the WWW, and consisting of WANs, MANs (metropolitan area networks), and LANs (local area networks) performing various application-specific functions.

Dave Dooling (top) and Rui de Figuerido (bottom)

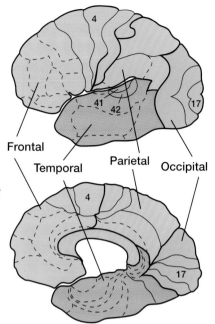

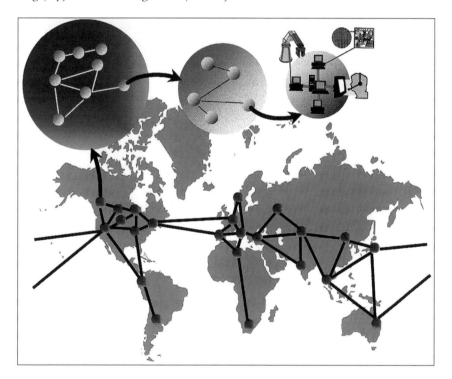

works (WANs) consisting of metropolitan area networks (MANs); these, in turn, consisting of local area networks (LANs), that ultimately link individual processors (computers and human/computer pairs). Its links (protocols) are optimized with regard only to speed and efficiency of point-to-point message passing without paying attention to the intelligence present in the communications.

What de Figueiredo anticipates in the twenty-first century is the Internet's evolution into a set of application-specific *functional* hierarchical networks, con-

RUI J. P. DE FIGUEIREDO: How can we accurately evaluate creativity and diversity?

"Our cultural, religious, and ethnic diversity can be turned into an asset if we develop an understanding of how to harmonize our differences to our advantage," stated Rui de Figueiredo. De Figueiredo is professor of electrical and computer engineering and of mathematics at the University of California at Irvine, and director of the university's Laboratory for Machine Intelligence and Neural and Soft Computing.

De Figueiredo's own eclectic background gives him an unusual perspective in bridging the gulf between C.P. Snow's famous "two cultures" of the humanities and the sciences. He was born in Goa, India, then an overseas territory of Portugal, and lived there until pursuing his bachelor's and master's degrees in electrical engineering at the Massachusetts Institute of Technology and his doctorate in applied mathematics at Harvard University. His educational background includes a Licenciateship in Music from the Trinity College of Music in London.

"I'm constantly on tenure committees, where people are trying to judge the different ways in which faculty members make significant contributions to their various fields,"

figured as neural networks running on top of its hierarchical anatomy. "These developments will occur as the Internet's functional architecture and orientation evolve from data service to multimedia," he said.

Now, a neural network—whether it be artificial (as in a computer) or biological (as in the brain) is a system of many rather simple processors (neurons) connected by links (synapses). The synapses are *active* links that perform functions on (assign weights to) the signals that pass through them. Thus, the input to each neuron consists of a weighted combination of output signals from a set of other neurons. If this input signal is higher than some threshold, the neuron fires—that

de Figueiredo continued. In this context, de Figueiredo's own experiences have led him to distill out the conclusion that "all creativity has two components: discovery and invention."

He explained: "Discovery is the finding of new things. It is creativity based on analysis. Physics, chemistry, and the other sciences are propelled primarily by analysis and discovery—for example, think of the work of Einstein.

> **'Discovery is . . . creativity based on analysis [while] invention . . . is creativity based on synthesis.'**

"Invention, on the other hand, is the putting together of known things in completely new ways. It is creativity based on synthesis. Engineering is propelled primarily by synthesis and invention—for example, think of the work of Edison and Shockley," de Figueiredo said.

"Now, the two are related, of course," he acknowledged. "Engineering, which is the epitome of invention, needs a foundation in the analysis and discovery of physics on which to build. Shockley is an embodiment of this approach in his invention of the transistor."

But to de Figueiredo, the more exciting aspect of this recognition of the universal components of creativity is the fact that "it's the beginning of a common language," he said. With it, "you can see what makes all disciplines hold together, and you have a way to ascertain the quality and value of people's contributions."

The recognition of different aspects of creativity also reveals when standards of judgment may be misapplied. For example, de Figueiredo noted, "a big problem in the engineering community is that people are judged on the basis of their analytical contributions instead of on their inventiveness."

In addition, viewing creativity as having components of both discovery (analysis) and invention (synthesis) allows one to see commonalities across fields that are usually

is, it produces an output signal that passes through synapses to a set of some other neurons. The process continues until a specific function (such as detection, classification, interpretation) is collectively performed by the many participating neurons, de Figueiredo explained.

"The neural structures in the human brain and in artificial neural networks naturally and mathematically lend themselves to be organized not only anatomically but also functionally in the form of hierarchical networks in performing functions characteristic of intelligence," he said. As an example, he pointed to how the human cerebral cortex is divided into 52 Brodmann regions, each of which—

viewed as disparate. For example, de Figueiredo pointed out, "with music and literature and the fine arts, you're looking more at invention rather than discovery. Thus, you see that engineering has more creative overlap with music and art than it does with pure science."

In the larger context of society as a whole, de Figueiredo feels that the lack of a competent "language" for political and social issues accounts for "our inability to articulate properly the concept of affirmative action," he said. "This country [the United States] is driven by a Constitution that is based on individual needs and rights without giving weighty and lucid consideration to societal needs and rights," he explained. "This has led to misunderstanding and conflict at the societal level as, for example, in the case of gun control legislation. Maximizing benefits to all people as well as to individuals requires compromise from both sides.

"Operations research is a field where optimization issues are being investigated. Life is a constrained optimization problem—a compromise," de Figueiredo noted, "just as the Heisenberg uncertainty principle describes an unbreakable compromise in physics between knowledge of a subatomic particle's momentum and placement."

◆

Rui J. P. de Figueiredo, who received his IEEE Fellow award in 1976 "for contributions to nonlinear system theory and to the application of spline functions to signal processing theory," served for a quarter century in the departments of electrical engineering and mathematical sciences at Rice University in Houston, Texas, where he led research in robots and automation for the National Aeronautics & Space Administration's space station Freedom. *He joined the University of California at Irvine in 1990, where his current research interests focus on nonlinear signal and image processing, neural networks, and machine intelligence.*

either solely or in cooperation with other regions—constitutes a hierarchical network of neurons that performs a function such as, for example, the higher-level visual or auditory processing. "One may view the Brodmann regions as the WANs of the human brain neural network," he said.

According to de Figueiredo, intelligence is inherent to a hierarchical network. "I define intelligence as the ability to use knowledge to solve problems. Another characteristic of intelligent systems, natural or artificial, is that they possess the ability to adapt, learn and/or evolve with experience." And that is what de Figueiredo sees emerging with the Internet revolution—the merging of local area networks,

Engineers and scientists rehearse procedures for an astronomy mission using the recently concluded U.S.-European Spacelab program. Many large organizations are studying the team-building methods used in the space program to mesh conflicting needs for scientific experiments on a mission. This room, for example, is just the mission management area. Several adjacent Science Operations Areas were customized by different science teams on each mission. *NASA/Marshall Space Flight Center*

intranets, the public switched telephone network, wireless communications networks, the World Wide Web, and other assets for processing and connectivity. These developments will arise from a dramatic enhancement of underlying systems' bandwidth and of an ability to store and identify large amounts (terabytes) of information in real time.

"I'm very much interested in a network's multimedia communications, and seamless integration of text, graphics, images, animation, sound, and video in human communications with networks, enhanced by multiple sensors and actuators." Multimedia communications is likely to lead to humans "being viewed as multi-sensory, multi-actuator objects," he suggested. "Thus far, computers and humans have lived as separate objects that interact through interfaces. But through multimedia and multiple sensors, humans and intelligent machines will be able to live in the same worlds and frameworks, working together seamlessly as intelligent, dynamical objects, without discrimination between living and non-living."

Does de Figueiredo think that ultimately a global hierarchical network of sensing, communicating, problem-solving processors and connections might attain enough complexity to become sentient, *á la* the lunar computer Mike in Robert A. Heinlein's classic 1964 science fiction novel *The Moon is a Harsh Mistress*?

"Certainly," de Figueiredo replied. "Sentience is the ability of a hierarchical network to recognize itself as being itself—as having its own 'one-ness' or identity distinct from that of other hierarchical networks living in the same environment. Sentience, or self-awareness, is an inherent property of a hierarchical network living in a world of hierarchical networks." [For another view on the potential sentience of complex computers in the twenty-first century, see the box "When computer intelligence exceeds human intelligence, what will it mean to be human?" by Ray Kurzweil in Chapter 7. And for de Figueiredo's own views on human creativity, see his box "How can we accurately evaluate creativity and diversity?" p. 74.] ◇

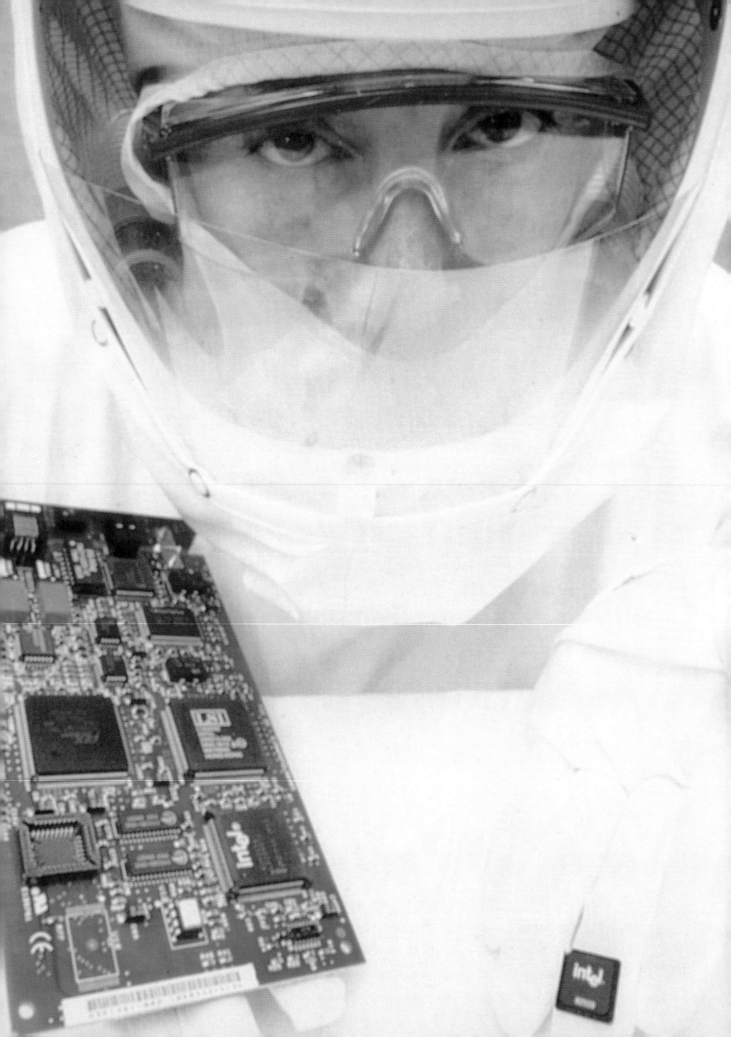

CHAPTER 4

COMPUTERS AND SOFTWARE

Faster, smaller, cheaper, and more powerful is the never-ending mantra for personal computers and software. Once the province of geeks and rocket scientists, computers now are everywhere, running everything, and being run by everyone. For this stroll through the silicon terrain, we called on two pioneers in personal-computer hardware and two in software.

Rao R. Tummala is Pettit Chair Professor in Electronics Packaging at the Georgia Institute of Technology in Atlanta and director of the university's Low-Cost Electronic Packaging Research Center. In his opinion, the key to continuing to reduce the cost of computer hardware is moving to a "system on a package." Gordon E. Moore, immortalized for the law (explained presently) bearing his name, is co-founder and chairman emeritus of Intel Corp. in Santa Clara, California, the company that developed the microprocessor. He outlines his perspectives on how fundamental quantum limits may ultimately limit the density of transistors in microprocessors, thereby limiting the continued growth of computer power. Moore believes such limits, however, may actually help improve software.

Left: An Intel technician in Hillsboro, Oregon, gives a graphic demonstration of Moore's Law of ever-smaller and more powerful chips. On his fingertip at right is the new single-chip solution Intel 82559 Ethernet controller, taking the place of the components on the board at left. The 82559 is 75 percent smaller and requires 75 percent less power than any previous controller. It also offers a single access solution for all the components in network-server, desktop and mobile computers. *Intel*

Making the computer "user friendly"—itself a computer-age term—is the province of software. One pioneer was Alan Kay, co-inventor of the graphical user interface (GUI)—such as today's familiar overlapping windows and icons—and of the concept of the personal and laptop computers. Now vice president of research and development at Walt Disney Imagineering in Glendale, California, he believes that software will not attain its full potential until the best software is available to all. And Douglas C. Engelbart, founder and president of the Bootstrap Institute in Fremont, California, was the inventor of the computer mouse, the first major step towards letting humans tell the computer what they want it to do. He explains why he feels that both the GUI and the mouse are now *limiting* future options for realizing the full potential of personal computers.

Efficient Hardware Packaging

"The current trend coming out loud and clear from the semiconductor houses is that of a system on a chip," said Georgia Tech's Rao Tummala. "Everyone is trying to integrate as much as possible on a single chip. The plan of large corporations is to end up with one wafer-sized integrated circuit—for a personal computer, cellular phone, camcorder, what-have-you—that encompasses the entire system."

At first glance, the concept of integrating an entire system on a chip seems attractive for two main reasons, Tummala explained. First, it eliminates much of the manual or automated assembly now involved in surface-mounting separate components (central processing unit, graphics chips, DRAM memory, SRAM cache, power supplies, resistors and capacitors, etc.) onto each printed-circuit board. Currently, even a simple cellular phone or pager may consist of 300 or 400 separate components. If they could all be manufactured as one chip, all the devices could be powered,

The inside of Apple's latest G3 Macintosh looks quite similar to the inside of its early '80s Apple II (or the inside of most other desktop computers): a motherboard plus slots for value-added boards and memory. But each of the chips delivers more power and performance with each generation. Ultimately, chips will be packaged to incorporate even more capability and reduce time lost communicating outside each chip. *Apple Computer Inc.*

Increased use of the Internet means increased demands for network reliability. Lineene Krasnow, IBM vice president, demonstrates how a rack of Netfinity servers can be rebooted from the palm of her hand with a WorkPad PC Companion connected to the Internet. The new technology allows the handheld device to diagnose and resolve network problems from a remote location. *Feature Photo Service and IBM*

cooled, and communicate with one another as a system. Second, a system on a chip minimizes the number of times a signal has to be routed off one chip and onto another—eliminating the resulting inevitable delays and reduced power. That translates into better performance at a lower price. "And the only way we can continue to reduce the cost of personal computers is to make sure the hardware gets cheaper," Tummala said. [And that accents his question, "How can we prevent ourselves from drowning in high-tech waste?" on page 98.]

But there also "are significant problems with designing an entire system to one chip," he continued. The physical chip may become so large that the signal's propagation time from one end of the chip to the other becomes too long, because the on-chip signal delay is, in general, longer than the off-chip delay. Thus, the benefit of a single chip solution can be lost at high speeds. Longer on-chip delays result from larger on-chip wire parasitics (that is, from the fact that the resistance and capacitance of the on-chip wire are larger than the off-chip wire).

Moreover, the manufacturing yield (ratio of perfect chips to total chips) of such complex wafer-sized chips becomes lower with increasing size—to the point where the cost of the chips becomes very high because the cost of processing a wafer is fixed. Also, wafer-scale integration fabrication plants cost U.S.$3 to $5 billion apiece, more than typical microprocessor-chip fabrication plants. Worst of all for a fast-paced market, flexibility is limited for updating the system once it is cast into silicon, and the time to market is longer than for ordinary subsystem chips.

For these and other reasons, Tummala and his colleagues at the National Science Foundation-supported Low-Cost Electronic Packaging Research Center at the Georgia Institute of Technology are advocating an intermediate concept they call a "system on a package." The package, called a single-level integrated module, or SLIM, integrates everything but the processor and memory integrated circuits which are placed on the package and connected, Tummala explained. Specifically,

DOUGLAS C. ENGELBART: Can we make society smarter?

"PHOTOTIME PORTRAITS" COMPLIMENTS OF BOOTSTRAP INSTITUTE

"There is a significant potential for boosting what I call the collective IQ," the intelligence quotient of human organizations, said Douglas C. Engelbart, the man who brought the human race one of those fundamental, seemingly obvious inventions that changes society: the computer mouse. He is also founder and president of the Bootstrap Institute in Fremont, California, an organization dedicated to helping society improve itself, one group at a time, through what he calls "improvement communities."

In Engelbart's perspective, the pursuit of knowledge—although done by individuals—is fundamentally a collective activity. "Community is exceptionally important," he observed. "I think we ought to have a saying: 'The organization is the real end user.'" He views organizations as social organisms that interact with the outside world with quantifiable aspects of intelligence. For example, Engelbart noted, one can ask: "How well does it make sense of what is going on, and anticipate, and predict? How well does it then make some kind of plan about what it should do about threats or opportunities? How well does it marshal its resources? How sensitive is it to the fact that there are changes, so now we need to modify our plan? To me that is something you can really call a collective IQ."

it incorporates all the infrastructure—the dielectric conductors, resistors, capacitors, inductors, and power supplies—on which any electronic system depends. The processor and memory chips that form the heart of the system, however, are manufactured separately and placed in the package.

Although there is still some assembly required, it may involve only three or four elements instead of 300 to 400, Tummala pointed out. But the flexibility provided by having several elements instead of one is considerable, he said. Paramount in importance, "the system-on-a-package approach gives manufacturers a platform for inserting new modules and new functions," he said, "allowing them

Improvement communities "are consortia of profession-al societies that are collectively trying to improve something: the knowledge for this, the cure for that," Engelbart said. "If we start focusing energy on boosting the improvement community's collective IQ, that's going to help it pull up the rest of society."

Of particular interest to the Bootstrap Institute are the developing nations. "You can look at a nation and say there actually is some *de facto* improvement infrastructure. Well, why not start taking a serious look and see if it's possible to improve the capability of that infrastructure? That is what we call bootstrapping, the origin of the name for our insti-tute. The better a nation or other organization gets at learning how to do collective knowl-edge work, the better it is going to get at improving all kinds of things." In this way, nations can get better at building their own roads and communications systems—learning how to fish instead of being handed fish. For the same motivations, Engelbart is also keen-ly interested in helping poorer or less technically apt school systems in the United States.

"The complexity and urgency of the world's problems are just going to snowball. And unless we can do something effective about improving our collective abilities to deal with complex, urgent things, humanity is doomed. Humanity has never, ever had to deal with such complex and rapid change. So we have no history of how to accommodate it. If we just go on what used to be the way of incorporating improvements, it's not going to work."

In his opinion, the world needs "a very different mode for [organizations to] accom-

> **'If we start focusing energy on boosting the improvement community's collective IQ, that's going to help it pull up the rest of society.'**

Carol Bartz, chairman and CEO of Autodesk, greets viewers from an Autodesk-designed "virtual set" while videotaping the broadcast introduction of AutoCAD 2000 in March 1999. Once the province of Hollywood effects artists, such tools are coming into broader corporate—and eventually community—use to boost corporate IQs. *PR NewsFoto & Autodesk*

'Unless we can do something effective about improving our collective abilities to deal with complex, urgent things, humanity is doomed.'

modate and improve ways of thinking and working. We just can't keep plugging in all kinds of new technologies" and raising the bar on expected output and productivity. Current techniques of collective learning are like "telling a person 'you are going to have to learn more and get stronger, but you can't spend any time exercising or going to classes because everything has got to be done right soon.' There would be no way a human being could do any changing. So there are some very serious issues" in organizations and society at large.

He paused reflectively. "I think it is a race. Either the complexity will snarl us up and we'll crash—or we'll get smarter about how we handle things collectively. There is both a threat and opportunity."

◆

Douglas C. Engelbart is director of the Bootstrap Institute in Palo Alto, California. His electronics career started in World War II, where he served as a radar technician for the U.S. Navy. After receiving his B.S. in Electrical Engineering at Oregon State University in 1948, he earned his doctorate in Electrical Engineering in 1955 at the University of California at Berkeley. He did much of his pioneering computer work in the 1960s and 1970s at Xerox Palo Alto Research Center. In 1999, he was awarded the IEEE's Von Neumann Medal for his contributions to computer science, including developing the "NLS" (oNLine System) that incorporated the mouse, display editing, windows, cross-file editing, idea/outline processing, hypermedia, and groupware. He also initiated the ARPANET Network Information Center (NIC).

to change and update products without having to redesign the entire chip—thereby reducing time to market for new products."

Limits to Computer Power?

Moore's Law, first enunciated by Intel's Gordon Moore in 1965, predicts that computing power will double every two years as engineers become smarter and more adept at making smaller chips with ever smaller components. Actually, Moore recounted, "in the initial Moore's Law, I had the complexity doubling every year. That worked for the first 10 years, which was all I was predicting. Since then, it has

The potentials of greater hardware and software complexity and power—good and bad—are illustrated by the terminally confused HAL 9000 in the 1968 movie, "2001: a space odyssey." HAL's icon status is so entrenched that in early 1999, Apple Computer used him to explain that the "Y2K problem wasn't our fault. It was a bug, Dave." *Apple Computer Inc.*

been doubling every two years" because the design process has become so efficient that "we no longer have the ability to tack things on by clever inventions."

Each time the density of transistors has doubled, the cost has plummeted. "As long as we were making circuits out of individual transistors, the cost was on the order of a dollar per transistor. We now sell transistors for less than a *millionth* of a dollar," Moore observed.

But, he cautioned, "obviously, anything that changes exponentially eventually runs into problems. The physical world doesn't allow for infinite exponential expansion"—or, in the case of computer chips, infinite contraction. Ultimately, Moore expects that the doubling rate of computer power will slow from two years to four years when quantum limits fundamentally reduce engineers' ability to make things smaller. "In several more computer generations, we will get to the point where we are approaching the atomic nature of matter and materials stop behaving like they do in bulk," Moore explained. "That will happen within the next 20 years. And then the rate at which things change will slow down."

Far from seeing the twenty-first-century day when software will end up saturating the capabilities of the chips, however, Moore thinks that a few hardware limitations actually might prove to be good for progress.

"Somebody once said that Intel giveth and Microsoft taketh away," he laughed. "Having more processing power has let software people be a lot looser in the way they write software" than they could afford to be in the early days of computers when memory and processing power were scarce and valuable. "Since memory is cheap and processing power is relatively cheap, the net result is that software can take a path that is maybe less effective than it potentially could be." In short, eventual limits to the growth of computing power could have the beneficial effect of imposing more rigorous mental discipline on software design to increase the efficient use of what power already exists.

Style is back in style. Apple Computer's iMac—widely believed to stand for Internet Macintosh—debuted in mid-1998 and revived flagging computer sales, providing impressive performance in an all-in-one package that other companies were moved to emulate. *Apple Computer Inc.*

Quantum limits to chip densities are "not the end of progress," Moore concluded. In the twenty-first century, computing "will still be moving at a rate that is unmatched by any other technology. We will be putting literally billions of transistors on a logic device. That leaves tremendous flexibility for the designers. I don't think it is going to be a real slow-down in the rate at which technology impacts society." [See his thoughts about "Are computers really the tide that will float all boats?" on page 94.]

Improving Software and Operating Systems

Another expert who believes that software is hardly as good or as efficient as it could be is Alan Kay of Disney Imagineering. Every day, users of Apple Macintosh and Microsoft Windows operating systems struggle with differences in languages and formats between their different types of computers, their operating systems, and various applications. Worst of all is the statistically all-too-frequent "blue screen of death" that appears when the Windows operating system crashes.

"Expert programmers have gotten mad enough [at commercial operating systems] to say to vendors, 'You people are completely bumbling along here,'" Kay declared. After years of being a niche player for scientific and other limited applications, UNIX—the terse, powerful operating system developed by AT&T Bell

Laboratories in the 1970s and later adopted by Sun Microsystems—is growing in popularity as a real-time, nearly "crash-proof" operating system. From UNIX have come other versions such as Linux, which is reputed to be crash-proof.

"Linux came out of [Finnish programmer] Linus Torvald's saying, 'This is not to be endured. I'll write a better one,'" Kay said. In his opinion, Linux is one "of the best operating systems available now." It can be downloaded free of charge from various sources on the Internet, although neither UNIX or Linux has as many applications as Macintosh or Windows. [See his thoughts on "How can we separate the Internet's wheat from its chaff?" on page 88.]

The struggle for dominance by Macintosh, Windows, and UNIX/Linux operating systems is mirrored by the struggle for dominance among software used for applications on the Internet. As the Internet grows over the next decade and hundreds of millions—perhaps even a billion—new units are connected, Kay sees no winner rising from the cyberdust. "There will be many different object-oriented hosts [moving blocks of code as if they were individual, physical blocks that can't be altered], so no one system—Java, C++, Squeak—needs to win out. None will win out because there always will be factions that are interested in each for one reason or another. It is hard to do a perfect system. But because they are object-oriented, they should be able to send messages from one object to another," something that is imperfectly implemented in systems today.

"Open Sesame." On March 31, 1998, Netscape Communications Corp. CEO James Barksdale released the source code—the "secret recipe"—for the firm's popular Navigator Web browser. This was the first time that source code for a major commercial product has been released to the public. Companies increasingly are looking to making source code available for little or no fee, but requiring a royalty for sales of "value-added" products derived from the code. *PR NewsFoto and Netscape*

ALAN KAY: How can we separate the Internet's wheat from its chaff?

"When you have a new publishing medium that has great extent, the easiest kind of content is low content," said Alan Kay, vice president of research and development for Walt Disney Imagineering in Glendale, California, who co-invented the graphical user interface (GUI)—today's familiar "desktop" and "windows." Just as there is so much "junk" available on off-air and cable television and on newsstands, he sees the same thing happening with the Internet. But he also sees hope for rich content—assuming users can get at it.

"The Internet, when it started in 1969, was all about sending essays and ideas from scientist to scientist," Kay said. "Then [in the 1990s] it broke out into the public as a popular and available thing. But probably 99 percent of the content on the Internet today is distractive, of no consequence." Kay expects that like the gold rush that opened the American West in 1849, only a few nuggets will emerge from the tons of cyber-silt. "But they will be as important as Newton's *Principia*," he predicted, referring to Sir Isaac Newton's seventeenth-century ground-breaking rewrite of basic physics.

Kay feels that part of the reason such a high fraction of published or televised con-

Kay sees two solutions for this imperfect world. The longer-term solution is open-source software, where authors reveal their software's source code to other programmers. "You see open-source software done by people who honestly want software to be better," Kay predicted. The process is not for the faint of heart because "open source" will also mean "open criticism." Software designers will have to quit working like alchemists of yore who kept their "magick" to themselves, and instead become like today's scientists, who put everything out for peer review. "People grudgingly realized they would make a lot more progress if they let other people debug their ideas, on the grounds that even wonderful scientists

tent is "junk" is because of marketing and the perception of public demand by companies and advertisers that hold the purse strings. "Marketing people basically go after people as they are," he observed. "Generally, they are not trying to improve people by the experience. They just want to sell them something. Many marketing people do not want people to get more discerning. Television could have good stuff on it. The limitation on the number of channels and the cost to produce programming, though, essentially preclude making programming good." Similar economics affect print media, he feels.

In Kay's opinion, the Internet's economics—specifically, its low cost of entry—and the equal worldwide access to any user makes it very different. On the Internet, there is a growing—but not yet complete—compendium of material on science. "This is going to be fantastic before another decade is out," he said. Similarly, "the Berkeley Literature Library is essentially a compendium of worthwhile writing about ideas from the past. You can find 15 different annotated versions of Plato's *Republic*, which, from the stand-

At once boon and drawback,
the Internet allows anyone to contact anyone, thus prompting parents to
worry. May 1999 saw the introduction of the NetSafety program designed to protect
kids from online predators and give parents the tools to monitor and control their
children's Internet content as well as have a direct link to law enforcement. Dev Toor,
12, right, and other sixth graders from Simon Baruch Middle School in New York City
experiment with the program. *Feature Photo Service and Imagebox*

'Probably 99 percent of the content on the Internet today is distractive, of no consequence.'

point of classical education, is really great."

Most important to users, "a lot of the stuff will be free, because marketing people don't believe they can sell high content." But that free availability could have profound consequences—especially for economically disadvantaged users—because "a lot of this educational stuff is going to start seriously competing with colleges." In effect, it will become possible for everyone to learn from Nobel laureate Leon Lederman," who won the Nobel Prize for physics in 1988 for his work on the neutrino beam method and the demonstration of the doublet structure of the leptons (very light particles) through the discovery of the muon neutrino.

"Lederman wrote *The God Particle* and *From Quark to Cosmos* and is now doing a lot of work in Chicago inner-city public schools," Kay continued. "If he gets [angry] enough [about available high-school science textbooks], he will write something himself, and this time it will go on the Internet," Kay said. "It will form a nucleus for other people to write similar high-quality stuff for the public good for free. And that will all of a sudden start competing very seriously with current high school textbooks."

Because the Internet will feature "some of the most important content that human minds have ever thought up," Kay feels that "obviously one of the most interesting problems is to make it as easy for people to find out about the good content as it is to find out about the distracting stuff that will be backed by billions of dollars worth of marketing. [Otherwise,] there is the possibility of the good content getting lost, as it is, indeed, in many public libraries. In any library, the percentage of books that are actually worth read-

are not likely to be as tough on the possible drawbacks of their ideas as they are in trying to show that their ideas are true," Kay said. "So science partially makes its way by giving glory and awards to people who find flaws."

In the meantime, consulting businesses are being established to help companies and individuals sort out confusing and incompatible operating systems and applications. "For a number of years [in the early and mid 1990s], the worldwide market for primary software from Microsoft and other companies was about $100 billion a year. In that same period, the worldwide marketplace in consulting—handholding and outsourcing—was $200 billion a year," observed Kay. "The latest

'A lot of the educational stuff will be free, because marketing people don't believe they can sell high content ... [but it] is going to start seriously competing with colleges.'

ing is probably around 5 percent." Yet, to the uninitiated user, nothing tells which 5 percent is worthwhile: "Every item in the card catalog looks like any other item."

In his view, the browser war of the late 1990s between Netscape and Microsoft's Internet Explorer will be remembered as but a minor skirmish that will fade when a new application evolves to help people find information based on content. "You could imagine a piece of software, a smart object, whose job it is to be a kind of editorial point of view on great content," to help users separate the wheat from the chaff, Kay said.

"Civilization is actually a learned skill," Kay said. "And it is not taught very well."

◆

Alan Kay, whose early work in 1968 on the Dynabook pioneered the concept of the laptop computer, was a founder of the Xerox Palo Alto Research Center (PARC) and a pioneer in artificial intelligence at Apple Computer (1984–86). He was elected to the National Academy of Engineering in 1997 for "inventing the concept of portable personal computing."

survey [late 1998] shows the consulting/outsourcing market is around $300 billion a year, whereas the primary software market is holding steady or—if anything— diminishing." What that means, he pointed out, is that "most need to buy help.

"The next interesting stage is what is being done by companies such as Red Hat, Inc. [of Durham, North Carolina]," Kay continued. Basically, they "are taking free software from the Internet and positioning their company as a packager and consultant. And this is the right thing to do," he feels, because it will increase the accessibility of superior software that may be somewhat less known, such as Linux, Squeak, and others. "And if open-source systems actually make their way into larger

A peak inside the production of *Engineering Tomorrow* reveals a computer mimic of a stack of papers. For computer power to be fully recognized, Kay, Engelbart, and others argue, software designers must think out of the box— or in this case, off the sheet of paper—and design innovative interfaces. *Dave Dooling*

markets through consultants that are basically paid to help companies achieve their goals, that is going to be amazingly powerful."

Of Mice and Men

For most users, the personal computer is symbolized by the graphical user interface (GUI) with its desktop and windows that are navigated by using the computer mouse. Ironically, many in the computer industry feel that the rapid, widespread adoption of the GUI and the mouse actually has stymied further development of the man-machine interface. "I've been saying for 25 years that people's picture of GUI just has to start evolving," declared Douglas Engelbart of the Bootstrap Institute. [He asks "Can we make society smarter?" with GUI and other techniques on page 82.]

In his opinion, the famous WYSIWYG—What You See Is What You Get— display is actually a step sideways. "I said in the '70s, 'That is neat—if all you want to do is emulate paper,'" Engelbart recalled. "But the computer can offer lots of different views and flexibility. Let's explore those, not just get stuck with emulating paper."

The current mouse-GUI combination—designed in the days when bandwidth and computer power were still at a premium—requires very little bandwidth for communication from user to computer: moving the mouse changes the X and Y coordinates of the cursor on the screen, and the finger clicks a button to act on the item where it rests. But there are additional "alternatives you can merge in with that," Engelbart said, to increase a user's speed and efficiency.

Specifically, Engelbart recalled a one-handed five-button mouse he developed back in the '60s, where the buttons could be pressed in combination like keys on a piano, to produce a total of 31 multiple-button chords. So, instead of first highlighting a paragraph ("a noun") and then going to a menu or to the keyboard to indicate that the text should be copied ("a verb") and then pasted ("another verb") elsewhere, a user would press the "copy" chord while highlighting the text and the "paste" chord when clicking at the destination.

"The freedom that immediately gives you is that when you go and click on something, your hand could have specified not only the noun but also the verb," Engelbart said. "If you hold down one of the mouse buttons when you strike a chord, you can make a character upper case rather than lower, or choose between punctuation and a numeral, or between a CTRL command or an ALT command. It just opened up the world immensely" in terms of speed and efficiency, he said.

And how about creating an interface so that users could feel as though they were reaching into the artificial world and interacting directly with what is "there," like Alice stepping through the Looking Glass? Although data gloves with position sensors and virtual reality headsets for three-dimensional viewing have been available for several years, processors and software have not yet advanced enough to make data gloves, sensors, and headsets widely available or affordable. Yet, their potential value is clear. In one application, for example, the U.S. National Aeronautics and Space Administration (NASA) is experimenting with using data gloves to let aerodynamics engineers poke their hands into the "air flow" around a virtual model so they can manipulate the design.

In the future, Engelbart sees a need for multiple classes of user interfaces ranging from pedestrian to high performance, with open-source standards to al-

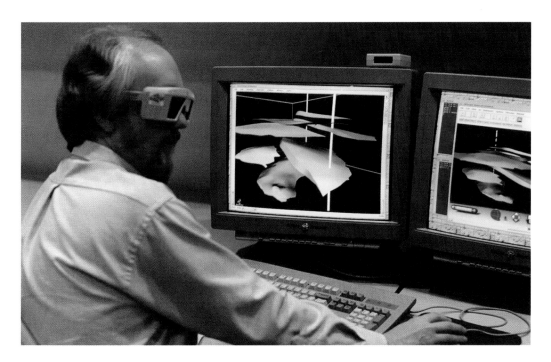

Virtual reality allows engineers and scientists to get "inside" anything from the inner ear to the Earth itself. At the Amoco Technology Center in Tulsa, Oklahoma, an Amoco geoscientist views three-dimensional models of a petroleum reservoir in South America. Amoco employees—including geologists, geophysicists, engineers and drillers—use the 3D models to gain a better understanding of reservoirs. *PR NewsFoto and Amoco*

GORDON E. MOORE: Are computers really the tide that will float all boats?

LOUIS FABIAN BACHRACH ©

"I am concerned that we could end up with a society that becomes too class-conscious based on education," cautioned Gordon E. Moore, chairman emeritus and co-founder of Intel Corp. in Santa Clara, California. One conventional wisdom about the economy is that when the tide rises—when prosperity spreads—it floats all boats. Yet the increased dependence of the industrial world on computers at all levels makes prosperity more like a tsunami, easily carrying those already navigating deep waters, but swamping and drowning anyone still in the shallows or waiting on the beach.

"In some respects we are seeing that already," Moore continued. "Look at a company like Intel. Twenty years ago we hired a good many assembly line workers based on their mechanical dexterity. We would test them to see if they could line up one pattern with another one easily under a microscope. We did not care what kind of education they had, really; not even what language they spoke.

low a choice of easy-to-learn tools. "There is no reason to think that the way we have of representing knowledge with the symbols and languages [we have] is the best way, just because that was the way we could with the printing [press]," he declared. Similarly, "the computer offers so many more symbolic options in helping [users] realize and portray concepts."

Predictions: Quantum Computing

One of the more speculative, but exciting, fields of promise is quantum computing, said A. Robert Calderbank, vice president of information sciences research

"In order to do those equivalent jobs now, they have to be able to use a computer, they have to have pretty good English-language skills. We hire almost nobody in our manufacturing plant who doesn't have at least two years of college. If Intel's requirements are indicative of the way the workforce is changing, there just aren't going to be many good jobs for people without some college. A lot of people could be left out" of the computer age's promised prosperity, he said. While Intel's philanthropic foundation donates equipment and money to schools and communities, it is only one company, and its contributions are the proverbial drop in the bucket compared to the national or worldwide need. "If Intel has to drag almost the whole population along, we have a major problem," Moore added.

Part of the problem is the inaccessibility of the computer as a device to its user. In *Citizen Soldier*, his history of American men fighting across France from D-Day to VE Day, Stephen Ambrose credits a large share of America's World War II victory to legions of shade tree mechanics who, once drafted, quickly adapted from repairing cars at home to repairing tanks and other weapons. They returned home to drive the postwar economy.

Similarly, until the 1970s, a practical electronics education—still based on electronic components—was largely accessible to any interested individual through Heathkits for assembling radios, stereo amplifiers, and other consumer products. Even the Altair, the first personal computer, was a do-it-yourself job that taught the basics to many of today's computer greats.

But today's computers are different. "Mechanical things are more or less intuitive," Moore said. "You can figure out how two gears

Zoe Potts, seven, of Portland, Oregon, explores the Disney Web site from her computer while her mother uses the same Internet connection to pay bills from a PC in the den. Their computers are connected by Intel's AnyPoint Home Networking products. These use existing phone lines to allow multiple-PC families to share Internet access, printers, files and multi-player games. *Feature Photo Service and Intel*

go together, how one turns the other. You can see what holds things together." Thus, decades ago "people without a formal education could participate in the automotive-dominant community. They could become car mechanics, they could do a lot of things without much [formal] education. But a real concern that I have about computers is that they are not intuitive. If you want to work on a computer, you can't just open it up and figure out how it works," Moore reflected. "As we progress further in the electronic age, it seems to me that we are more dependent on some kind of formal education to participate."

That worries him. "For the U.S., it requires a major national thrust to truly address educational problems," he said. "We may have to go through a generation of people who don't participate [in high-tech prosperity] to get the message across that education is the only way to participate successfully.

"I wish I had an easier solution. But a lot of it is getting the message to the right people that education is the only way to participate."

◆

Gordon E. Moore, who received his IEEE Fellow award in 1968 "for contributions and leadership in research, development and production of silicon transistors and monolithic integrated circuits," is president emeritus and co-founder of Intel Corp. In 1997, he was awarded the IEEE Founder's Medal for his "world leadership in very large scale integration, and for pioneering contributions in integrated circuit technology." He also was awarded the 1990 National Medal of Technology by President George Bush, the Founders Award from the National Academy of Engineering, and other IEEE honors, including the W.W. McDowell Award from the IEEE Computer Society, the IEEE Frederick Philips Award, the Computer Pioneer Medal and the IEEE John Fritz Medal.

at AT&T Labs in Florham Park, New Jersey. Pioneered by AT&T Labs colleague Peter W. Shor, who won the International Congress of Mathematicians' coveted Nevalinna Prize for distinguished work by young (under age 40) computer scientists, quantum computing harnesses the quantum mechanical properties of individual atoms to explore an extraordinary number of possibilities at the same time.

"Shor showed that if you build a quantum computer, you could factor integers [positive whole numbers] very fast," Calderbank explained, referring to the process of finding the smallest prime numbers that, when multiplied together, yield the original integer. "This is important, because many public key cryptographic

As computer users store more information, storage technologies struggle to keep up with the demands for reliability and fast access. Quinta's optically-assisted Winchester (OAW) technology can store the equivalent of 45 copies of the 32-volume *Encyclopedia Britannica* within the space of a postage stamp. "Winchester" harkens back to the earliest days of desktop computers: It was the code name for IBM's first hard drive (10 megabits!) for PCs. *PR NewsFoto and Quinta*

schemes rely on the presumed intractability [extraordinarily time-consuming difficulty] of factoring large numbers for their security."

Ordinary bits in today's digital computers are binary: they have only two states—0 and 1, or off and on. In the counterintuitive quantum-mechanical world of the atomically small, however, we see quantum bits ("qubits") where the underlying physical system is in both the 0 *and* 1 state at the same time. When measured, the system collapses to either 0 *or* 1.

"It's a very difficult notion," Calderbank acknowledged. "Even Einstein couldn't come to grips with it. He didn't like indeterminancy of quantum-mechanical probability—in fact, that's the origin of his famous comment that 'God does not throw dice.'

"But because this superposition represents the atom's being in many states simultaneously, it allows you to process exponentially many things at the same time," Calderbank explained. Thus, the quantum-mechanical properties of atoms could make powerful parallel computation possible. "And in 1994, Shor figured out how to use this model to make factoring easy."

In principal, a quantum computer could be very reliable. "We showed how you could build a classical reliable computer out of unreliable components using error correction," Calderbank recounted, referring to fundamental work he published with Shor in 1996. "The question was whether you could also do that in the quantum world. People thought 'no' because it might violate the Heisenberg uncertainty principle," he said, referring to the physical law discovered by Werner Heisenberg early in the twentieth century. Heisenberg's uncertainty principle states that it is impossible to know both the precise position and the precise momentum of a subatomic particle at the same time, because the act of observing the particle

RAO R. TUMMALA: How can we prevent ourselves from drowning in high-tech waste?

GEORGIA INSTITUTE OF TECHNOLOGY

"The single biggest issue facing society in the twenty-first century is the preservation of the environment," brooded Rao R. Tummala, Pettit Chair Professor in Electronics Packaging at the Georgia Institute of Technology in Atlanta and director of the university's Low-Cost Electronic Packaging Research Center. "Western high-tech disposable culture is so hardware-driven and is putting so much chemical and physical waste into the environment that, as a country, the United States is not doing anywhere near an adequate job of waste disposal, let alone a good job.

"In Japan, I get the question 'what are you doing for the environment?' more often than in any other country," continued Tummala, who splits his time between Georgia Tech and the University of Tokyo. "Ecodesign" is a watchword among Japanese engineers, who are focusing on reducing harmful materials in waste by a factor of 1,000 in the early twenty-first century.

"Japanese engineers *design* a product from day one with disposal in mind," Tumma-

changes one condition or the other. "But by new understanding of something called quantum entanglement, we discovered you can indeed do error correction in the quantum-mechanical world," Calderbank continued. "So, if you can create quantum components, you should be able to build a reliable quantum computer."

So far, however, a quantum computer remains a twinkle in theorists' eyes, because no real one has yet been built. The difficulty is primarily practical. "If you want to factor the number 15 [that is, find 3 and 5], you'd have to produce a quantum computer with between 20 and 30 bits—that is, you'd have to manipulate the energy states of between 20 and 30 individual atoms," Calderbank said.

la explained. Personal computers and other high-tech products are being designed with "dry mechanical connections" that don't leak toxic materials when disconnected so that "at the end of their life, the user can disassemble them, returning some parts for reuse or recycling and placing others for safe environmental disposal."

Japanese high-tech industries are reducing the use of lead-based solders in electrical and electronic products—in fact, such solders will be completely prohibited as of 2002. Moreover, new manufacturing processes are being devised to minimize chemical waste. A principal target is the process by which printed-circuit boards (PCBs) are made. "For every gram of copper that ends up in a PCB today, 10 to 20 grams are wasted," Tummala said. "The current technology calls for depositing copper over the whole board and then etching away 95 percent of it to leave the lines for electrical connections. The Japanese, however, are looking at processes for depositing only the 5 percent of the copper that you want." There is also an active search for manufac-

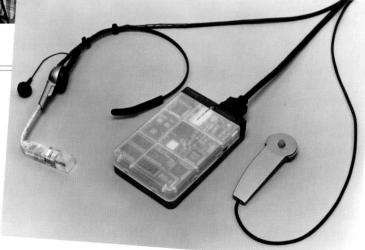

Smaller, wearable computers may be one step in reducing high-tech waste. IBM is among several organizations experimenting with "wearable" computers that will have a tiny display screen and voice recognition software. Such machines will be employed first in industrial settings for calling up technical information, but no doubt will migrate outwards to consumers who revel in technology for its own sake, and to being plugged into the 'Net non-stop. *IBM*

'We need not only to design for manufacturing and design to cost, but also to design for disposal.'

turing materials that are soluble in water instead of ones that require toxic organic solvents.

In designing products, all engineers "need to look systematically at the *whole* life cycle of the product," said Tummala. "We need not only to design for manufacturing and design to cost, but also to design for disposal with minimal environmental impact."

In Tummala's view, rampant unconcern about technology's environmental impact is a global issue. In Europe and the United States, "companies are not much interested [in ecodesign] because their job is making money, not saving the environment. Sure, the U.S. is a big country with a lot of space, but that's no reason to destroy it," he said. "We should be a model for ecodesign and environmental responsibility instead of a follower."

The prognosis for developing nations is even grimmer. "Already, China and India are the two worst countries in the world in terms of the destruction of their environments," due to uncontrolled emissions from motor vehicles and smokestack industries, he said. "Pollution from China has shown up as far away as Singapore, Taiwan, and even all

In July 1997, Norfolk (Virginia) International Airport became the first airport in the world to provide business travelers with CyberFlyer, a new public Internet access booth created by CyberFlyer Technologies. Travelers can now take a seat in the next generation phone booth and, with the swipe of a credit card, surf the World Wide Web and check e-mail, flight reservations, and airline schedules as well as access a variety of other online services. *PR NewsFoto and CyberFlyer Technologies*

the way to California," he pointed out. "You can imagine what will happen as they keep growing in electronics manufacturing. Unless those nations can self-impose discipline—externally imposed discipline never works—they will end up killing themselves, drowning in their high-tech waste."

He paused thoughtfully. "The question really is: does high technology help humankind? People think 'absolutely,' because of the prosperity it brings. But if you sit back and think what it's doing to people's lives, the environment, and the quality of life, it's clear the effects are not always positive."

> **'The question really is: does high technology *help* humankind? Its effects are not always positive.'**

◆

Rao R. Tummala, who earned his Fellow award in 1994 for "contributions to the development of multichip packages for high-performance computers," was born in India, where he lived until receiving his two undergraduate degrees (B.E. in metallurgical engineering from the Indian Institute of Science in Bangalore and B.S. in physics, mathematics, and chemistry at Loyola College's Indian campus in Nungambakkam, Chennai). Before joining Georgia Tech in 1993, he was an IBM Fellow at the IBM Corp., where he invented a number of major technologies for IBM's products for displaying, printing, magnetic storage, and multichip packaging. He is coeditor with Eugene J. Rymaszewski of the widely used Microelectronics Packaging Handbook *(N.P.: Chapman & Hall, 1997), the author of 90 technical papers, and the holder of 21 U.S. patents.*

"At the moment, physicists can manipulate the states of only one or two atoms at a time," he added, citing the work of John Preskill at the California Institute of Technology in Pasadena. "The challenge is, can we realize the mathematics of quantum superposition in macroscopic bulk systems? We're not sure it's going to scale. So, there's a gap" between theory and practicality.

Nonetheless, quantum computing is real. "Real physical systems at an atomic scale are quantum systems that we can play with," Calderbank said. "And physicists are wonderfully inventive. I hope we'll find a physical framework for quantum computing. And I am sure we'll see remarkable things in the next decade." ◇

CHAPTER 5

COMMUNICATIONS

"ANYWHERE, ANYTIME" COMMUNICATIONS is the dominant vision for the twenty-first century, as expressed by many of the industry's practitioners. To that might be added "any type," for the ultimate vision is that of a wireless handheld terminal that a person would carry everywhere, and which is capable of receiving everything from phone calls to e-mail to television programming to electronic books, and can allow its user two-way Internet access.

To make that vision reality, many businesses that have been traditionally regarded as separate media have been getting their hands into one another's technologies and markets over the last decade. Wireline telephone companies are buying up or partnering with cable television companies and wireless communications companies to offer TV and Web services over coaxial cable, optical fibers, and wireless channels. Cable TV companies and Internet service providers are competing with each other and the traditional telephone companies to offer voice telephony as well as Web access. Magazine publishers are experimenting with "Webzines" while book publishers and retailers are investing in manufacturers of electronic book (e-book) reader devices and services that download e-books through the Internet. And every traditional communications industry—telephony, television, radio, and publishing—is looking over both shoulders at the encroachment

Left: Globalstar (shown) and other satellite-based systems will put all the corners of the earth, and virtually all income levels, within communications range of the rest of the planet. *Globalstar*

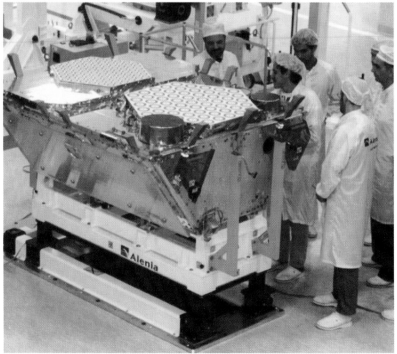

A gateway—ground station—for Globalstar "large LEO" satellite communications system located in South Korea (top) and a Globalstar communications satellite is assembled in Rome (right). Four or more Globalstars at a time can be launched from a single rocket. The white polygons on top are the phased-array (i.e., electronically steered) antennas that connect individual users to the gateways. Solar arrays deploy to the left and right. *Globalstar*

of all the aggressive broadband competitors, nervous that something might be gaining on it.

What form those technologies will ultimately take and which industries will dominate communications in the twenty-first century, however, are up for grabs. "People talk about 'convergence' as if communications and computing are coming together into a center," remarked Bennett Z. Kobb, president of New Signals Press

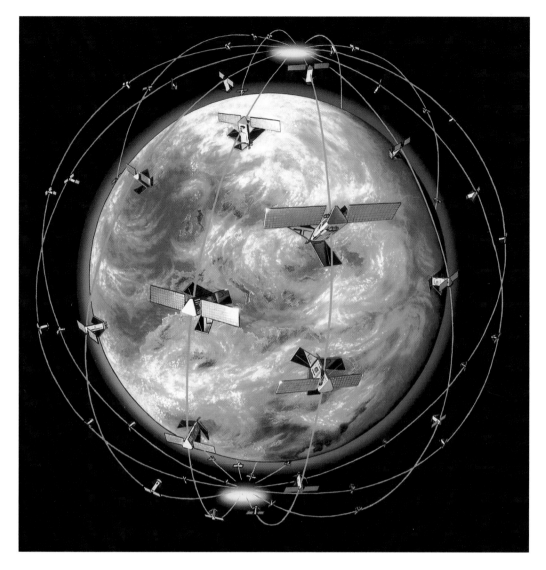

Globe depicts how low Earth orbit communications satellites will cover the world. Satellites will hand off to each other as they rise and fall above the horizon so the user's call continues uninterrupted. *Iridium*

in Arlington, Virginia. Kobb, a veteran observer of wireless communications and U.S. Federal Communications Commission (FCC) rulings for technical publications, is a co-founder of the Wireless Information Networks Forum, a wireless-communications industry trade group. "I see it as businesses splashing around into each other's former territories."

No one, however, doubts that the Internet is going to be the primary driver. "The growth of Internet traffic in the long-distance arena is presently doubling about every 12 months while voice traffic is increasing around 10 percent per year," observed Tingye Li. Li, a communications consultant in Boulder, Colorado, spent more than four decades at AT&T Bell Laboratories (and, in the last two years before his retirement in 1998, at AT&T Laboratories–Research). "We expect that as soon as

TINGYE LI: Are we eating our seed corn?

"Because of competition, high-tech industries—particularly in the United States—cannot afford to support basic research as they did 20, 30, or 40 years ago. So companies like Lucent, AT&T, IBM, and others now mainly support applied research with a vision that is relatively short-range," said Tingye Li, former Division Manager at AT&T Laboratories–Research. "Professors in physics departments of many universities are also doing more applied than basic research.

"But what knowledge will we have to depend on 10 years from now? The applications we're deploying now depend on fundamental discoveries that were made fully 20 or more years ago. But industry is too occupied with its quarterly bottom line and competition in the marketplace that it doesn't have the time or resources to devote to fundamental strategic research. And you don't see that type of work being supported in universities now.

"To be sure, there's still room for growth and enhancement and innovative ideas in current technology.

"For a while.

2001 or 2002, data traffic will overtake voice traffic on long-distance trunks"—and then keep on accelerating, both in developed and developing nations.

In other words, "instead of a voice-based telephone network, we are moving to a data-based network," said Rod C. Alferness, chief technical officer for the optical networking group at Lucent Technologies Inc. in Holmdel, New Jersey. Moreover, the explosive demand for Internet access is "coming at the same time as the potential for an enormous increase in bandwidth" on optical fibers.

Probably within the twenty-first century's first decade, "we will reach a point where all communications traffic is packet-switched, although packets associated

"But for us to open *new* fields in the decades ahead, we need to support fundamental research—the type Bell Laboratories used to do in the 1950s, '60s, and '70s and earlier," Li declared.

"That leaves only governments. Governments can stimulate and support strategic research—basic science that is focused on specific areas of application. If successful, it would open a wealth of new fields for the next century.

"But this requires people in government with vision, who understand the real needs of industry. Unfortunately, in the United States, senators and congressional representatives worry more about short-term international competition. So more government dollars go to support near-term applied research—the very type that industry ought to be doing itself.

"So, what institution is now going to pioneer fundamental research to fuel tomorrow's technology?"

'**For us to open *new* fields in the next century, [this country needs] to support fundamental research.'**

◆

Tingye Li, who received his IEEE Fellow award in 1972 "for contributions to laser resonator theory and the application of digital computation to the analysis of lasers and microwave antennas," retired in 1998 after 41 years with AT&T (at first, Bell Laboratories, later AT&T Laboratories–Research), doing research on lightwave technologies and systems. He and his colleagues pioneered research on optical-amplifier and wavelength-division multiplexing and championed their application.

with real-time services, such as voice, may be singled out for special handling," added A. Robert Calderbank, vice president of Information Sciences Research at AT&T Laboratories in Florham Park, New Jersey. And, in his view, all this data will not be carried only by optical fibers: "what's going to transform [local access] communications in the twenty-first century is wireless data."

Optical Superhighway

The exploding demand for Internet access is single-handedly driving a powerful technology on long-haul fiber-optic trunks: wavelength-division multiplex-

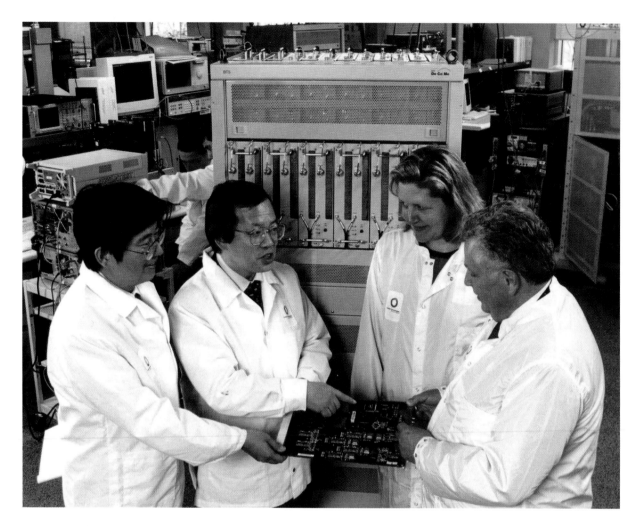

Technical executives from NTT DoCoMo and Lucent Technologies inspect Lucent's prototype W-CDMA base station in Mt. Olive, New Jersey. *PR NewsFoto and Lucent Technologies Inc.*

ing (WDM)—the ability to combine onto one optical fiber many different channels, each an optical signal at a slightly different wavelength or "color."

Silica optical fibers are among the most transparent media on earth. Moreover, the theoretical bandwidth of silica optical fibers is many terahertz, meaning that theoretically it should be possible for thousands of high-speed (gigabits per second) channels to share a single optical fiber.

Despite this dazzling bandwidth, for the last quarter-century the practical capacity of an optical fiber has been limited to a single channel because of one major technological and economic barrier: reliable broadband amplifiers. Since the 1970s, the most reliable form of amplification has been optoelectronic regeneration, in which the optical signal is received by a photodetector that converts it to an electronic signal. The electronic signal is then amplified and regenerated, and then used to modulate a semiconductor laser that emits an exact replica of the

Optical fibers made of ZBLAN (right)—fluorides of zirconium, boron, lanthanum, antimony, and sodium—hold tremendous potential for communications across the near-ultraviolet to infrared spectrum. The challenge is that production on Earth yields defects that distort the signal. Tests indicate that processing in space could yield defect-free fibers. Dennis Tucker (below) at NASA's Marshall Space Flight Center is developing techniques to verify ZBLAN properties. ZBLAN fibers could be produced aboard the International Space Station. Medical applications, remote sensing, and optical amplifiers are just some of the benefits that ZBLAN may yield. *NASA/Marshall Space Flight Center*

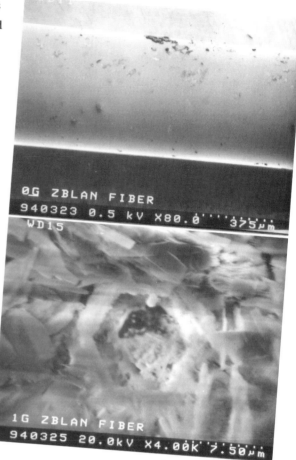

original optical signal. This process requires every optical signal to have its own dedicated fiber with its own set of regeneration equipment. Thus, until the 1990s, the primary focus of many communications engineers was on increasing the bit rate of each fiber's individual channel rather than on increasing the number of channels per fiber.

In the late 1980s, the Holy Grail of a broadband optical amplifier was discovered in the form of ordinary silica optical fiber doped with atoms of the rare-earth element erbium. When a strand of this erbium-doped fiber only a few meters long is bathed in the continuous illumination of a semiconductor pump laser, electrons in the erbium atoms are raised to metastable higher energy levels. When photons of a weak optical signal enter the fiber, they trigger the cascade of some of the erbium electrons down to their ground state. Each cascading erbium elec-

ROD C. ALFERNESS: When is unlimited information effectively no information at all?

COURTESY OF LUCENT TECHNOLOGIES

"Through the Internet and the Web, we soon will have worldwide access to all the information that's ever been generated," said Rod C. Alferness, chief technical officer of the optical networking group at Lucent Technologies Inc. in Holmdel, New Jersey. "But how can we organize and present that information in a way that's meaningful and efficient for the human mind to absorb? And how can we know we can *trust* it?"

Alferness' first concern is sheer quantity of information. "In this information age, when in principle all information is available to everyone in huge storage banks and brought to us through high-bandwidth pipes, the question now is: how can we *reduce* the amount of information to that which is truly meaningful and useful to our needs?" In his view, the sheer volume is approaching the point where individuals can become so

tron emits a photon—at exactly the same wavelength and phase as that of the entering signal.

In other words, a humble length of erbium-doped optical fiber is an extraordinarily efficient broadband amplifier—and right at the infrared wavelength region (1.53 to 1.60 micrometers) of silica fiber's minimum attenuation (maximum transparency). Even better, when multiple optical signals at slightly different wavelengths enter the erbium-doped fiber simultaneously, all are simultaneously amplified—each at its own wavelength, phase, and bit rate. Best of all, the wavelength region over which the fiber amplifies is some 8 terahertz wide. In theory, that means

swamped that they can't process it, effectively putting them into the same predicament as if they had no information at all. After a certain point, the larger question is: do you have half a lifetime to look at everything?" he asked. "And since many documents repeat others, you must ask—as I do with television programming—whether enormous bandwidth means that you have *content*?"

Alferness' second concern is quality of information—of telling the wheat from the chaff. "This is one of the major issues facing professional societies," he noted. "What is unique about professional journals? They are refereed—there is a review process that attempts to authenticate the validity of the information." But with electronic publishing and the ease of any individual "just putting it up on the Web," he said, a publisher no longer acts as a natural gatekeeper to ensure that only quality makes it into print. "Will that mean that the peer review process will eventually go away? And will that, in turn, jeopardize the level of authenticity of information available on the Web?" Alferness asked.

Even for information that would not normally be subject to peer review, quality and authenticity is an issue, he pointed out. For example, while in many consumer magazines there is a clear distinction between objective editorial material and advertising, on the Web, sometimes that distinction is less clear. In fact, some material that poses as objective information may actually have been written and placed by advertisers of certain products. Moreover, many kinds of electronic information are posted in formats that can be manipulated, so they "are so easy to pass on to other users and potentially may be corrupted" by an intermediary before doing so, Alferness said.

> **'You must ask—as I do with television programming— whether enormous bandwidth means that you have *content*?'**

---

one erbium-doped fiber amplifier (EDFA) can simultaneously amplify close to 800 optical channels, each operating at 10 gigabits per second (although engineering constraints limit present commercial systems to "only" about 100 channels at 2.5 gigabits per second).

By the late 1990s, in what Tingye Li characterizes as "the largest upgrade of the telecommunications network in the history of mankind," EDFAs were retrofitted to fiber cables buried in the United States 10 to 20 years earlier. "Almost overnight, each fiber already in the ground was turned into the equivalent of 8, 16, and now 80 using wavelength-division multiplexing (WDM)," said Li—and at a

'Is something on a disk truly archival? Do electronic media have a physical lifetime longer than paper?'

Alferness is also concerned about the information's physical durability. "Another special aspect of professional journals is that they are archival: once they are published, they are maintained forever," he said. But with electronic publishing, many professional societies and libraries are rushing to discard everything on paper. "Yes, paper eventually disintegrates. But is something on a disk truly archival? Do electronic media have a physical lifetime longer than paper? Can we trust the information not to disappear over time?" he asked. Moreover, electronic formats evolve far more rapidly than books—much information on old computer punch cards or tapes is now permanently inaccessible (if it were not converted to some other medium) because there are no more punch card or tape readers.

In his opinion, information overload, validation, and durability are issues that engineers, Internet service providers, and Web site designers must address "*early* in the twenty-first century."

◆

Rod C. Alferness, who received his Fellow award in 1989 "for contributions to integrated optics and guided-wave devices, in particular for the conception and demonstration of tunable waveguide filters and of high-speed photonic switches and modulators," is the co-author of five book chapters on optical waveguide structures and devices.

cost far cheaper than the U.S. $60,000 per kilometer entailed in installing all-new fiber cables. [For Li's opinions about whether more such fundamental inventions lie ahead, see his box "Are we eating our seed corn?" p. 106.]

Other countries are quickly following the United States' lead—in fact, China's long-haul optical-fiber network of more than 80,000 kilometers now surpasses that of AT&T's in the United States. Moreover, much of it is WDM—not only because of its superior capacity, but also because EDFAs are far cheaper to deploy than optoelectronic regenerators.

But EDFAs have done far more than just cost-effectively open the door to allow

WDM to take full advantage of the raw capacity of optical fiber's bandwidth, said Lucent's Rod Alferness. WDM also offers "a means of true multinode networking" essential for handling Internet traffic.

Traditional point-to-point voice telephony has relied on circuit switching: setting up a physical electronic circuit that is dedicated to each telephone conversation for as long as the call is in session. But the Internet Protocol (IP)—which is optimized for data networking—relies not on switching but on routing: data streams are divided into individual packets whose headers have the destination's address. Each packet then finds its own individual route through the network, being reunited in proper order with its neighbors after it arrives at the destination. Packets may be of any length or bit rate.

To Alferness, the most exciting promise of WDM is that it "allows you to multiplex different bit rates onto the same fiber," he said. "You can even use different data formats on the same fiber—such as SONET for voice conversations on one channel and IP for data transmissions on another." [For Alferness' view on the quantity and quality of information on the Internet, see his box "When is unlimited information effectively no information at all?" p. 110.]

Wireless Everything

Within the first decade of the twenty-first century, one wireless pocket terminal that can access either terrestrial or satellite facilities to transmit images, text, Internet connectivity, computing power, and other services is anticipated to be operational anywhere in the world.

Dubbed "superphone" by the British daily newspaper *The Financial Times*, the International Telecommunication Union's official name for this 3G (third generation) technology is IMT-2000 (International Mobile Telecommunications 2000)—although it is also known in Europe as Universal Mobile Telecommunications Services (UMTS). It is called "third generation" because it is the third major incarnation of wireless telephony after the first and second generations of cellular and personal communications services (PCS), respectively. Aficionados of wireless history may also recognize that IMT-2000 was once called the Future Public Land Mobile Telecommunications System (whose unwieldy acronym FPLMTS was variously pronounced "Flumpits" or "Flumps").

Operating at about 2 gigahertz, IMT-2000 is expected to offer high-quality voice transmission and short message services similar to those available from current PCS providers. In addition, it will offer wireless data communications similar to that now available through phone lines to people with a high-speed computer modem. The wireless data rate could eventually scale up to between 1 and 10

BENNETT Z. KOBB: The electromagnetic spectrum—public trust or pork barrel?

"The old days of start-up companies petitioning for radio frequencies from the U.S. Federal Communications Commission are over, except in novel and creative services and low-power unlicensed devices," brooded Bennett Z. Kobb, co-founder of the Wireless Information Networks Forum, an industry trade group. Kobb is a Washington D.C. technical journalist who has covered wireless technologies and the U.S. Federal Communications Commission for nearly two decades.

PAUL JAFFE

"Now that the U.S. Congress requires the FCC to auction everything (except digital television frequencies) when there are competing parties, the wireless field is slanted toward deep-pocket players and the subscriber model, in which consumers must pay for services rendered," Kobb said. Auctions of the electromagnetic spectrum were inaugurat-

megabytes per second for instant Web browsing, downloading of files, and even full-motion videoconferencing. Named after the year in which it was originally slated to be launched, IMT-2000 is expected to be a network of networks that will allow users to roam the globe with the same handset—unlike current cellular and PCS phones, whose frequencies do not match across all three of the International Telecommunication Union (ITU) regions of the world.

Regardless of the exact form of its ultimate incarnation, IMT-2000 is projected by some to "one day compete with—and perhaps even supplant—desktop computers, and conventional wired and wireless telephones," reported New Sig-

ed in 1994 as a source of revenue to reduce the U.S. Federal debt. Before then, the radio spectrum was viewed more as a public trust than as a Federal revenue source. Applicants for FCC licenses could obtain them without paying enormous sums at auctions, although license applicants still faced administrative, legal, and engineering fees."

As a result of the auctions, "uses that do not involve paid subscription services for the general public—such as private radio operated by oil fields or courier services—are not favored, and are finding it hard to obtain spectrum to serve growing communications needs. Moreover, the U.S. Congress requires the FCC to establish minimum bids, which can be very costly. So auction revenues are the driving force behind what services are being made available.

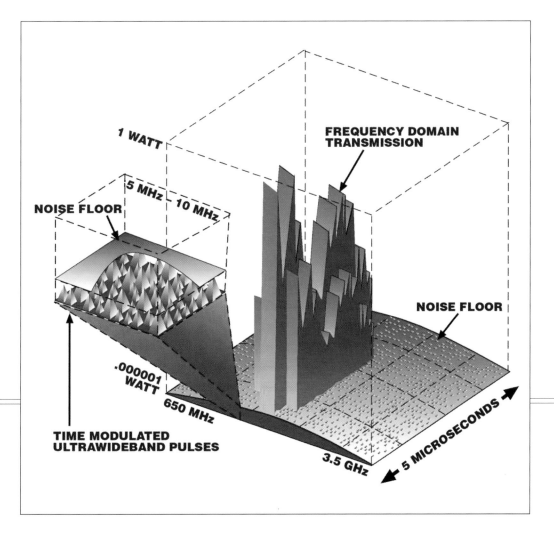

Crowding of the conventional radio spectrum may be alleviated by an unconventional technique: ultrawide, picopulse radio transmissions. In a new technique pioneered by Time Domain Corp., short pulses are sent out either ahead of a precise timing interval (1) or behind (0). The technique is not noticed by conventional radio, yet stands out above the radio noise "floor" for the appropriate receiver. Broad applications are anticipated in communications, radar, and medical and security imaging. *Time Domain Corp.*

'Auction revenues are the driving force behind what services are being made available.'

"The rule now is that if there are competing applications for spectrum, the winner cannot be decided by hearings or proposals or lotteries or first-come-first-served," as decisions were made for the first half-century after the FCC was created in 1934. Instead, Kobb explained, "money from both start-up and established companies that formerly could have gone to develop technologies or infrastructure or to pay employees now goes to the U.S. Treasury to buy a license. Some licenses have sold for billions of dollars, after which some license winners have gone bankrupt. Those licenses are now tied up in bankruptcy court. And there are severe penalties for non-payment of winning auction bids: if winners can't generate revenue to pay for the license over a specified period of time, they must actually pay *more*. Also, licenses can't be renewed unless FCC-established 'benchmarks' are satisfied; that is, there has to be some demonstration that the licenses were actually used to serve customers.

"The irony is, in frequency allocations, there are many engineering solutions that can be applied so that a service can be designed to allow many comers to get what they want. In some cases, *everyone* who wants it can be accommodated. But if plenty of spectrum is available because of ingenious engineering solutions, then the FCC may not get much money for the spectrum—just as you can't get money for the air that's available for everyone to breathe.

"The problem is, the FCC seems to be under pressure, both internally and from Congress, to design services so that there *are* mutually exclusive applicants—sometimes the FCC acts as if it is *required* to restrict access to spectrum so that its economic value shoots upward.

---

nals Press's Bennett Z. Kobb. Moreover, just as video "Netcasts," "Net radio stations," "e-books," and high-fidelity music are becoming widely available on the Internet, IMT-2000 or its successors could carry services that will compete with today's radio and television broadcasting, as well as interactive videogaming and print publishing. [For Kobb's concerns about the future of wireless services, see his box "The electromagnetic spectrum—public trust or pork barrel?" p. 114.]

The infrastructure behind superphones will include not only WDM optical fibers, but also communications satellites. These satellites are not the traditional "birds" in geostationary orbit (GEO) some 36,000 kilometers above the earth, whose

> **'But if plenty of spectrum is available because of ingenious engineering solutions, then the FCC may not get much money for the spectrum—just as you can't get money for the air that's available for everyone to breathe.'**

"Former FCC Commissioner Reed Hundt once commented that the initials FCC should stand for Federal Cash Cow," Kobb reflected. "The auction revenue has become a great way for the FCC to accumulate power and—in this era of increasing deregulation of communications services—to keep itself from being disbanded.

"But the auctions also mean that small, creative businesses are being left out. And I worry that the FCC is being turned into a merchandising agency."

◆

Bennett Z. Kobb, president of the consulting company New Signals Press, Inc. in Arlington, Virginia, has covered the FCC and wireless communications for IEEE Spectrum, Telecommunications Reports, *and the Voice of America's* Communications World. *His electronic sourcebook,* SpectrumGuide: Radio Frequency Allocations in the United States, 30 MHz–300GHz *(Arlington, Va.: New Signals Press, 1999), is now in its fifth edition.*

---

stately 23-hour, 56-minute orbit matches the earth's rotation to keep the satellite in one apparent position over land or sea. They will be fast-moving satellites in a low earth orbit (LEO) some 1,000 kilometers high (about twice the altitude of the highest orbit attained by the manned space shuttle). Because such low orbits mean that each individual satellite can't communicate with very much of the earth's surface at one time, dozens of satellites are arrayed in a constellation that scans all of the earth with satellite beams. Among other advantages, LEO satellites' closeness to the earth eliminates the disconcerting propagation delay due to the time required for the telephone signal to travel up to GEO and back to the earth at the

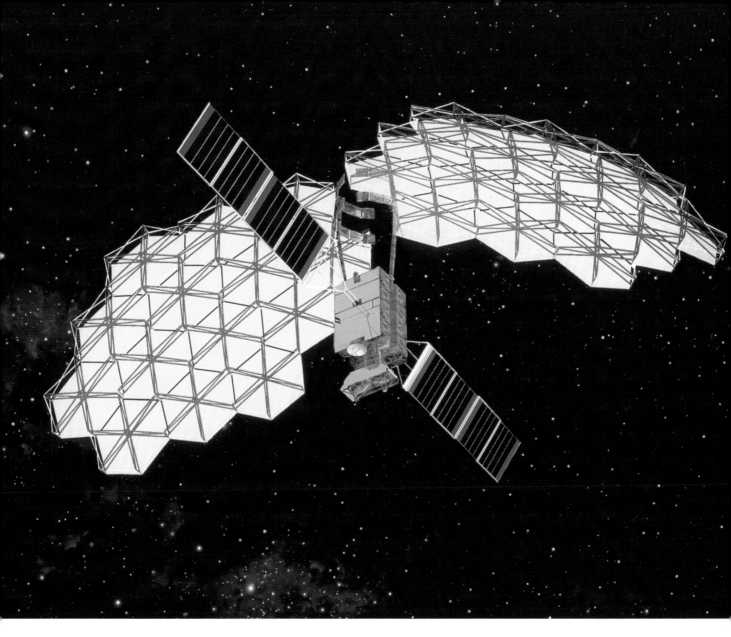

In 2001, Japan will launch the ETS-8 Engineering Test Satellite with a 20-meter (66-foot) deployable antenna to serve mobile communications needs. Large, deployable antennas will increasingly be used to provide tightly focused spot beams, some of them tailored to follow national borders, and to connect with low-power transmitters and receivers.
National Space Development Agency, Japan

speed of light. In fact, voice communications via satellites in LEO have a propagation delay that compares favorably with the imperceptible 35-millisecond delay over transoceanic optical fibers.

But crucial differences between data and voice communications pose knotty engineering challenges that must be resolved before high-speed Internet traffic can be carried on wireless channels. Voice communications are constant while data communications proceed in bursts. So to accommodate data, "we need to optimize the radio link and all network layers above it for bursty traffic," said AT&T Laboratories' A. Robert Calderbank.

"Wireless channels are nasty," he explained. "If you have only one transmitting antenna and one receiving antenna, it's very hard to send lots of data without lots of errors. If you want a high data rate on a wireless channel, you need multiple antennas for both transmitting and receiving. Obviously, you can't turn your Palm Pilot or cell phone into a hedgehog [porcupine], but you'll get close to 1 megabit per second on a 200-kilohertz channel if you have four transmitting antennas at the base station and two receivers on your terminal," he said. Thus, in his view, high-speed wireless data will become practical business in the twenty-first century when unobtrusive multiple antennas are developed with "new space-time coding across the antennas for reliability" at data rates comparable to those carried by coaxial cable or optical fibers. [For some of Calderbank's thoughts on future communications developments, see his box "What is the role of industrial research laboratories in the twenty-first century?" p. 120.]

Predictions: Spectrum Free-for-All?

"The conventional wisdom is that radio spectrum is scarce, must be fought over, bid for, and parceled out in little chunks," said Kobb. But an unusual wireless technology—ultrawideband radio (UWB), pioneered by patent-holder Larry Fullerton of Time Domain Corp. in Huntsville, Alabama—may be poised to set that assumption on its head.

UWB transmits radio signals in nanosecond-long "monocycles"—pulses so brief they occupy only a single cycle of a radio-frequency wave, Kobb explained.

As the cell phone industry explodes, providers are searching for new places to install relay antennas—including existing power structures. Sam Dellinger of Duke Engineering & Services Power Delivery Field Services works on top of one of the 34,000 existing towers ranging in height from 15 to 183 meters. Duke Communication Services is offering the wireless communications industry excellent alternatives to free-standing towers with single attachments. *PR NewsFoto and Duke Power*

A. ROBERT CALDERBANK: What is the role of industrial research laboratories in the twenty-first century?

PHOTO COURTESY OF A. ROBERT CALDERBANK

"The traditional view of big industrial research laboratories like [the Thomas J. Watson Research Center of] IBM or Bell Laboratories was that they provided a pipeline from fundamental science to prototypes to advanced development to—hopefully—commercial products," mused A. Robert Calderbank, vice president of Information Sciences Research at AT&T Laboratories in Florham Park, New Jersey. "It was a closed world, in which one company could pretty much determine a market and did not need to pay much attention to what was going on outside its walls.

"The landscape in which research takes place in communications technologies has fundamentally changed in recent years," Calderbank continued. "Today, it's seldom the case that one company makes a market. Now there's a tide of innovation in products and services that involves fierce competitors with a compelling interest in ensuring that their

The technology offers high communications privacy, low interference, low cost, and high consumer capacity, he said. Moreover, when used in radar applications it can also make extremely precise measurements of distance.

In principle, UWB is basically a logical extension of conventional spread spectrum technology, which is already used for code-division multiple access (CDMA) cellular telephones and short-range wireless data. But, Kobb notes, UWB also has one significant difference: While conventional spread spectrum signals require a few megahertz to about 20 or 30 megahertz of bandwidth, UWB uses vastly more spectrum—up to several gigahertz.

products and services interoperate. Thus, companies—and their industrial research labs—must face outward and influence the outside world, while still extending the scope and reach of the work to give the company a market advantage."

In an environment where competitors' devices must interoperate, standards move to front and center, Calderbank stated, "because standards make it possible for large companies to come to a common point of view without running afoul of antitrust considerations. This is *not* a statement that research distinction is becoming less important. Actually, it's becoming more important. But communications companies today must leverage it differently. They must use it to exert an influence on the standards activities in which competing companies coordinate their offerings. If your company has distinguished folk making technical arguments, then your technical arguments have more leverage than your competitors'.

"In this context, the goal of research activity is to seed a larger community, and to influence that community in a way that provides the right science and technology for you. Thus, companies must do a lot of networking, particularly with professors and students who are naturally focused on the kinds of problems that are important to you. Moreover, standards-setting is a political process: you can't ignore the points of view of the other participants—that's a recipe for failure.

"In such an environment, an individual company can no longer feel it's important to win all the way. What it should care about instead is simply that *somebody* provides a solution that is in line with its thinking. That perspective in itself is a significant difference from the past."

To Calderbank, the view that industrial research laboratories are no longer important or even necessary to a company is short-

Turning corners degrades signal strength and quality, so it is important to know what the physical layout of an optical fiber system may do to the data being transmitted. A technician measures signal loss for a laser beam transmitted through an optical fiber that is looped on a test stand.
Lucent Technologies Inc.

'The goal of research activities is to seed a larger community, to influence that community in a way that provides the right sort of science for you.'

sighted and inaccurate. Even though some companies have diminished, outsourced, or disbanded their labs, others—such as Microsoft—have started ones. "What was the big event for Microsoft?" he asked. "Being blindsided by Netscape and the advent of the Web browser. If Microsoft had had a research lab earlier, it might have awakened to the importance of the Internet and browsers. Companies need research labs in the sense that you get yourself a canary or an insurance policy: if you don't see the future, chances are that the future will knock you over the head.

"Moreover, if you have competitors, you're always looking to create reasons that you should do better than they do. The key is thinking insightfully and systematically about your basic business and bringing good ideas to it"—in short, doing research.

"Industrial research labs can and should do several interesting things," Calderbank asserted. "First, they're always going to have a role in exploring improbable propositions—things for which you can't yet make a business case but might be profitable in the future, such as real-time language translation. Second, they're going to be increasingly important in influencing standards-setting—because if no standard is widely accepted, the market gets captured by a proprietary language or standard and every company that depends on what has been captured loses its independence. Third, industrial research labs provide valuable institutional memory, learning from past successes—and failures.

"You may hear that industrial research labs have lost their position in visionary

Thus, Kobb pointed out, UWB wireless services would leap the boundaries of traditional spectrum allocations, unaffected by the intricate legal conditions that apply to each frequency band. Properly coded and operating at low power, UWB signals would appear as ambient radio noise to other users of the same spectrum—if it's noticed at all. FCC tests suggest that UWB should, in fact, have no significant impact on other users (although some engineers have expressed concern about the aggregate impact of large numbers of UWB devices across a nation).

The advent of UWB and other novel spectrum-efficient technologies has caused at least one law professor to predict that within the twenty-first century's first

research. I don't believe that. In certain areas and locales, such as cryptography in the United States, they're the *only* place to do unclassified research, because the [U.S.] Federal government won't fund cryptographic research in universities.

"If anything, universities themselves have become more conservative—less visionary—over the past 20 years because scientific disciplines have defined themselves more narrowly, and because the [U.S.] National Science Foundation has been under pressure to attach all kinds of performance metrics to research grants," Calderbank remarked.

"There's a difference between pioneers and real estate developers—between improving existing processes and creating a new process that is more chancy but possibly has a bigger ultimate payoff. At the moment, the National Science Foundation is not at all visionary in what it funds—it's backing real estate developers: you almost need to know how good your idea is before you can get a grant. DARPA [the Defense Advanced Research Projects Agency] has a much better record in funding pioneering research.

"You don't want to be religious about doing one and not the other—you want balance in the portfolio," said Calderbank. "Portfolio management is something individuals do all the time. And the best companies do it very well."

◆

A. Robert Calderbank, who received his IEEE Fellow award in 1998 "for contributions to the transmission and storage of digital data," is one of the inventors (with Nanbi Sashadri and Valud Torokh) of space-time codes, breakthrough wireless technology that uses a small number of antennas to provide superior data rates and reliability. He also directed AT&T's effort to develop the V.34 high-speed modem standard, and is a pioneer in quantum computing.

decade, some court would rule that spectrum regulation is unconstitutional, said Robert W. Lucky, corporate vice president of Telcordia (formerly Bellcore Applied Research) in Red Bank, New Jersey. "The idea behind his statement is so compelling that you start to believe it," Lucky remarked. "In theory, if transmitters obey certain courtesy requirements, there may be a limitless ocean of spectrum capacity, without interference. The tragedy of the commons may not be operating here—everyone can bring their sheep to graze without hogging the resource." And with infinite spectrum capacity, "no government may have the constitutional right to regulate it any more." ◇

CHAPTER 6

ENTERTAINMENT

ONE OF THE LARGEST TRENDS in entertainment media is wholesale digitization. A host of broadband services will soon render it impossible to distinguish television data from radio, and make both look like telephony and Internet access.

"In the United States, we are on a track to convert all television broadcasting from analog to digital," declared Wayne C. Luplow, executive director of digital business development and high-definition television for Zenith Electronics Corp. in Glenview, Illinois. "The Federal Communications Commission has suggested a deadline of 2006 to have analog transmitters turned off and all the spectrum space occupied by analog TV channels turned back over to the Federal Government."

Radio broadcasting is doing the same, in two incarnations: terrestrial digital audio radio services (DARS) and satellite digital audio radio services (SDARS). Satellites are also bringing digital radio to developing nations at the same time as to developed markets, said S. Joseph Campanella, who was vice president and chief scientist at Comsat Laboratories in Bethesda, Maryland, for 25 years and who is now chief technical officer for WorldSpace Corp. in Washington, D.C.

Left: Computers are playing increasingly large roles in movies, augmenting what actors and prop masters can do. A new range of possibilities was heralded in 1998 by a jester, of all "people," in "Tightrope," an award-winning five-minute short produced entirely in computers by Digital Domain, Inc. In the movie, the jester on a tightrope tries to make his way past a masked man. The characters are more than cartoons; they have textured skin and fluid movements that add to the realism. Special effects now encompass "seeing is believing" computer images that resurrect dinosaurs, the Titanic, and even the Titanic's lost passengers. *Digital Domain, Inc.*

Both television and radio are digitizing not only to increase fidelity and capacity, but also to position themselves as media for accessing the Internet and the Web. Among other things, the Internet may facilitate "group wisdom," ruminated Robert W. Lucky, corporate vice president of Telcordia (formerly Bellcore Applied Research) in Red Bank, New Jersey. And the desire for instant, continuous connection may lead to wearable computers, added Stephen B. Weinstein, a Fellow in NEC USA Inc.'s C&C Research Laboratory in Princeton, New Jersey.

Mandatory Digital Television

The digitization of TV broadcasting began for real in the United States in 1998, when a handful of stations began transmitting digital programs a few hours a day on an experimental basis. By the end of 1999, several hundred stations are scheduled to have gone digital, and by the end of 2000, it is slated to be a couple of thousand. And well before the end of the twenty-first century's first decade, all the U.S. transmitters that ushered in the age of television for its first 70 years will be silenced forever. Other countries are eagerly following the United States' lead to implement digital television as well.

Comparison of formats for standard- and high-definition TV (SDTV, HDTV). Boxes are drawn to indicate relative coverage in pixels assuming that the camera's pixels cover the same field of view on the scene being broadcast. Conventional NTSC (USA) video would be slightly smaller than the SDTV image at center. Both SDTV formats have the same physical screen size, but the 704 x 480-pixel image has slightly narrower pixels, and slightly greater horizontal resolution, than the 640 x 480-pixel image. *Dave Dooling*

Dallas Parcells of Sharp Electronics watches a display of HDTV sets at the 1998 Consumer Electronics Show in Las Vegas. While manufacturers promise new kinds of interactivity and more programming options, making it the hottest thing to hit TV since color, many consumers are updating singer Bruce Springsteen's complaint to say, "Five hundred channels and nothin' on." *Feature Photo Service and Zenith*

The digital TV system adopted by the FCC is a completely independent standard—named ATSC after the Advanced Television Systems Committee. Instead of burdening the new digital service with the 70-year-old, spectrum-inefficient and power-inefficient NTSC (National Television System Committee) system to ensure receiver compatibility, an approach called "spectrum compatible" is being followed, explained Wayne C. Luplow of Zenith Electronics Corp.

In the analog NTSC system, to avoid unacceptable interference that would result if every 6-megahertz television channel were assigned in a city, only every other channel is assigned at VHF frequencies and only every sixth channel at UHF frequencies. Such wide spacing, of course, results in a tremendous waste of valuable spectrum space. With the digital ATSC system, however, it is possible to insert a 6-megahertz digital signal into the currently unused slots between NTSC channels without interference either to or from existing NTSC transmissions. Thus, every current NTSC broadcaster has been assigned a second channel for the new, digital ATSC high-definition television (HDTV) service. In the future, when virtually all consumers have the capability to receive the ATSC signal, the analog transmitters will be turned off, and the spectrum space can be used for other services (not necessarily consumer television).

Although the overall ATSC digital standard and the new broadcast frequencies have been determined, there are still 18 possible scanning formats, which vary in resolution from today's screens to true HDTV. For the HDTV formats, the aspect ratio (ratio of viewing screen width to screen height) is 16:9 instead of 4:3, as in traditional televisions, so it is more like the silver screen in a movie theater.

WAYNE C. LUPLOW: What is the most environmentally sound way to dispose of consumer electronics products?

"As engineers, we need to be aware of the environment and the effects of high technology on it," mused Wayne C. Luplow, executive director of digital business development and high-definition television for Zenith Electronics Corp. in Glenview, Illinois. "Historically, at least some engineers have been aware that we must be stewards of the earth, because engineers are the ones who must find technological solutions. For example, when we found that certain types of insulators pollute groundwater, we had to

Why should viewers even consider paying the big bucks commanded by digital TVs over current analog sets? In a word, realism. The higher-resolution screen allows a viewer to sit much closer than is customary with conventional TV, giving rise to the illusion that the action is being watched through an open window and that people and objects are actually three-dimensional. "Once people see the picture and hear the sound, they'll buy them," Luplow confidently asserted. [Luplow, however, has concerns about the disposal of all the useless analog televisions, which he airs in his box "What is the most environmentally sound way to dispose of consumer electronics products?" above.]

find ways to eliminate the PCBs [polychlorinated biphenols] and ozone-depleting compounds out of the components." Engineers have also made headway in devising environmentally friendly processes for manufacturing components.

Now driven by the determination of the U.S. federal government to replace all existing analog TV broadcasting with digital TV broadcasting at completely different frequencies, a significant challenge facing engineers is figuring out "what is the most environmentally sound way to dispose of a television set," Luplow explained. "Today you just put it out at the curb and the junk

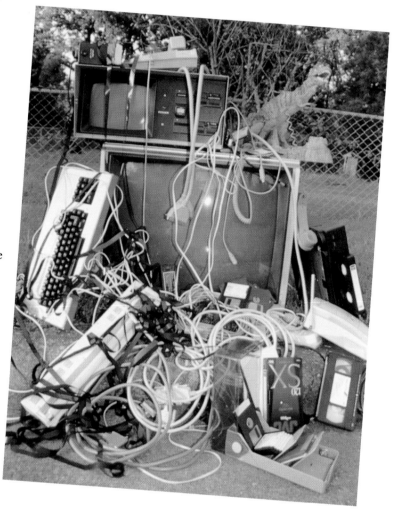

Electronics can be both boon and hindrance to the environment. The items waiting here for curbside pickup—all gleaned from one household and home office (with a few items left in the closet)—once were top-of-the-line electronics (with the exception of the dinosaur novelty phone). *Holly Dooling*

Terrestrial TV broadcasters are eager to implement ATSC because "it enables them to compete in a digital world" against cable TV operators, direct-to-home satellites, and videodisks. That is partly because of the stunning quality of the image: "on-screen noise and ghosts are things of the past, since multipath reception problems are eliminated by equalizers," said Luplow. But the biggest motivation is that "the digital TV system is a tremendous vehicle for the delivery of data by terrestrial broadcasting or cable." If even 18 of the 19 megabits per second in the channel are used for HDTV video and audio, there is still more than 1 megabit per second left for interactive data services—such as Web and Internet access.

'Most plastics in a TV set are not biodegradable, nor are the big pieces of glass used for picture tubes.'

guys take care of it. There's no special way of treating it, such as there is for turning in the lead core of a car battery." While he is skeptical that the FCC's aggressive schedule for phasing out existing analog TV frequencies by 2006 will proceed as planned, there is the potential that as the deadline approaches, "mounds of TVs could sit on the curbs."

His worry is that "most plastics in a TV set are not biodegradable, nor are the big pieces of glass used for picture tubes or any of the chemical elements used in TVs. So a TV lasts a long time in a landfill"—just like many other consumer electronics products, he added.

"And what about the pollution caused by the energy consumption in TV sets and other consumer electronics products? Or pollution of the radio spectrum by broadcasting stations?" In Luplow's opinion, finding solutions for these and other environmental issues pose "great opportunity for concerned engineers."

◆

Wayne C. Luplow, who received his IEEE Fellow award in 1996 "for leadership in establishing the terrestrial broadcast system of digital high definition television for North America," first joined Zenith Electronics Corp. in 1964 as a research engineer. He has headed Zenith's work in HDTV since 1993. He is the editor of IEEE's Transactions on Consumer Electronics.

"The convergence of TV, consumer electronics, communications, and computing is where all the fun is going to be," said Luplow. Digital TV technology opens the door to "putting Web browsers in TVs, TV receivers into PCs, and integrating everything with two-way communications for both playing and working."

Where might all the technological capability for distributed computing, Internet and Web access, and multimedia interactive wireless communications ultimately lead for both work and entertainment?

As a frequent business traveler who dislikes lugging a laptop and would like to secure the privacy of his viewing screen when seated in an aircraft, NEC's

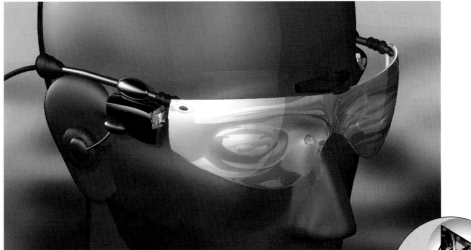

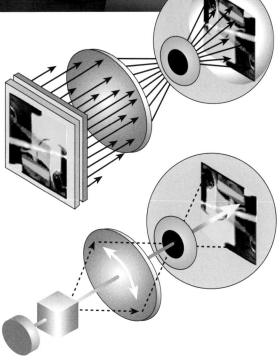

An advanced head-mounted display being developed by Microvision will use special low-power lasers shining into the viewer's eyes. Current head-mounted displays, such as Sony's Glasstron, rely on ultraminiature vidicons or LCDs to produce the image, which an optical system then projects into the viewer's eye (above). Microvision's Virtual Retinal Display will scan a low-power, multicolor laser directly onto the retina to paint the image there. In the video goggles above, mirrors take the place of the lens in this drawing. *Microvision*

Stephen B. Weinstein has his preferences. "My personal dream is a pocket-sized computer that has a microwave connection to a head-up see-through display on the inside of my prescription eyeglasses," said Weinstein, who researches software architecture, access networking, and multimedia systems and applications.

"Instead of a keyboard, I put on a lightweight data glove and type using the tray table of any airplane," Weinstein continued. "Through its wireless connection it can access local networking resources including Internet servers and the nearest printer, mixing and matching its capabilities depending on what's available." These components will, Weinstein believes, eventually become autonomous IP (Internet Protocol)-based devices, interacting with distant people and applications. [For Weinstein's views on how the Internet may connect humans rather than alienate them, see his box "How will the Internet affect social relationships?" p. 132.]

STEPHEN B. WEINSTEIN: How will the Internet affect social relationships?

"Unlike some people, who fear that the Internet is squelching human relationships, I think it is enhancing them," mused Stephen B. Weinstein, a Fellow in NEC USA Inc.'s C&C Research Laboratory in Princeton, New Jersey. "But it is also stressing them.

"One study done by Robert Kraut at Carnegie Mellon University in 1998 suggested that people who use the Internet a lot were more lonely. And a number of researchers have expressed concern about young people and adults spending so many hours in front of a computer screen surfing the Web, fearing that the Internet could have the effect of isolating human beings from one another.

"I doubt that the Internet makes people lonely. In fact, lonely people are probably using the Internet to reach out, to make themselves less lonely. It's been my observa-

'Color Radio'—No Longer a Joke

"Digital radio will offer better audio quality freed from interference—comparable to that of compact discs," declared Bennett Z. Kobb, a veteran Washington D.C. technical observer of the FCC and wireless applications. Digitizing the terrestrial radio spectrum also could increase capacity, "meaning that it could increase the number of entrants and opportunities for programming"—although, in fact, new entrants will have to bid for their licenses at auction, he said.

Currently, digital terrestrial radio broadcasting is being introduced both in the United States and Europe by the addition of digital sub-carriers to already

tion, in fact, that the Internet increases the number of people with whom you're in contact, and how often you interact with them. E-mail, for example, allows you to do what people do when they're physically proximate, ranging from planning conferences to sharing personal information and photos of their kids. I've also seen it let disabled people or elderly people who'd lost direct personal contact—say, with children who've moved away—regain part of that daily immediacy that is not felt through the mail or phone.

"But the pleasure of having a huge circle of contacts is complicated by the effort of maintaining that enlarged circle—and by the guilt of not succeeding. Responding to an e-mail message, for example, requires the physical act of writing, and I can spend an hour composing a reply because I want to think about what I have to say. But with e-mail there is an expectation that you'll get a quick answer, as you do in spoken conversation—and so you feel under pressure to respond instantaneously. Moreover, if a group of people in an e-mail discussion gets too large, the discussion can lose concurrency and can collapse in confusion as replies cross—just as if everyone were in a group talking at once, but not so easy to stop and correct.

"Hopefully, over time, social etiquette will develop around these electronic media, to relieve the pressure for an immediate response in e-mail and to moderate group discussions.

"There's also the issue of kids on the Internet—the social question of how much

Tak Yazawa of Sony's Entertainment Robot of America division and his real dog, Peanut, check out AIBO—Artificial Intelligence roBOt. The $2,500 AIBO, Japanese for "buddy," can walk, sit, shake hands, mimic emotions, and play with a ball. It contains a 64-bit RISC (reduced instruction set chip) processor, 16 megabytes of RAM, a CCD camera to follow the ball, and stores its programming on Sony's Memory Stick. In early June, the first 3,000 units sold out in 20 minutes over the Internet. *Feature Photo Service and Sony*

'Should there be restrictions on the exchange of disturbing material that is completely computer-generated without the exploitation of actual human beings?'

time a child or teen should be spending in front of a computer screen instead of playing outdoors face-to-face with other kids.

"And what about human rights? This same technology makes it possible for people to create what's in their imagination with graphic realism. So what happens when someone can create disturbing images without the exploitation of actual human beings? How should society react to the exchange of disturbing material that is 100-percent computer generated? What will it mean for personal relationships when highly realistic virtual people enter our lives?"

◆

Stephen B. Weinstein, who received his IEEE Fellow award in 1984 "for contributions to the theory and practice of voice-band data communications, and to IEEE publications activities," is a former president of the IEEE Communications Society. He researches communications software architecture, access networking, and multimedia systems and applications. He is the author of Getting the Picture: A Guide to CATV and the New Electronic Media *(Piscataway, N.J.: IEEE Press, 1986) and a co-author of the textbook* Data Communications Principles *(New York: Plenum, 1992).*

existing FM broadcasting carriers. Terrestrial radio broadcasters are also scrambling to digitize because they are "scared of two satellite companies—because people listening to satellite radio are not listening to local terrestrial broadcasts," Kobb pointed out. The two satellite companies are Satellite CD Radio Inc. in New York City, and XM Satellite Radio in Washington, D.C. At a spectrum auction in the late 1990s, both companies were awarded 12.5-megahertz chunks of radio spectrum around the frequency of 2.4 gigahertz to provide direct digital radio programming from satellites to automobiles.

Most profoundly, digitizing radio allows terrestrial broadcasters to provide

You can take it with you—at least in the car. Johnson Controls is marketing AutoVision, a vehicle-integrated entertainment system that enables rear-seat passengers to play video games and watch their favorite movies on videocassettes. By law, it has to stay in the back seat in the United States because it would be considered a distraction to the driver. *PRNewsFoto and Johnson Controls*

other services, including data communications and even images. "Digital radio broadcasting could transmit pictures—even, eventually, moving pictures," Kobb remarked. "So who's to say whether it's television or radio?"

Indeed, the word "radio" pales beside the broadband services actually envisioned for SDARS in the United States. "Direct digital broadcasting from satellites for *everything* is in the future," stated WorldSpace's S. Joseph Campanella. Satellite DARS systems will communicate satellite-direct digital broadcasts to small mobile and personal portable radios. Aside from music, the satellite worldwide digital broadcasts will include "static and dynamic images, text messages, paging, multimedia, and data," Campanella continued. "An example of direct data from satellite is the marriage of SDARS with GPS [the Global Positioning System]. Maps displayed on screens in cars and showing GPS locations will be updated in real time to indicate construction zones and other hazards. I fully believe that satellite DARS and GPS will be in every automobile in 10 years—that it will be a standard thing to have a dynamic electronic map along with music and other display information."

Meanwhile, WorldSpace's immediate aim is to use SDARS to bootstrap developing nations directly into the twenty-first century. "WorldSpace has authoriza-

S. JOSEPH CAMPANELLA:
What is the future of the U.S.'s universities and corporate research laboratories?

PHOTO PROVIDED BY WORLDSPACE CORPORATION

"If we keep destroying our technical institutions here in the United States, we're going to be second class," declared S. Joseph Campanella, chief technical officer of WorldSpace Corp. in Washington, D.C. Campanella, a pioneer of such key communications satellite technologies as echo cancellation, has headed research laboratories for much of his career. Among positions held during his 25-year career with Comsat Corp. in Bethesda, Maryland, he was vice president and chief scientist for Comsat Laboratories, developing fixed and mobile satellite technologies for the international Intelsat and Inmarsat systems. "I don't know where we're headed."

Some of Campanella's concerns center on one fundamental institution that has been instrumental in the development of twentieth-century engineers and technologies: universities.

"Educational standards are decaying in the nation," Campanella asserted. "When I had the opportunity to go to college—1945–50 for a B.S.E.E. and 1950–65 for an M.S.E.E. and Ph.D.—I thought it was a great and wonderful gift. It was my duty to take maximum advantage of the opportunity to learn. Today, too many students go to college to party rather than to get educated, with the attitude of challenging the professors to teach them.

tion from the International Telecommunication Union for direct audio broadcasting" of CD-quality sound from satellites to small radios on the earth, Campanella explained. The authorization "doesn't specifically say 'digital,' but digital is the best technique," he said—and so digital it is.

To gain the efficiencies needed to be cost-effective for direct digital delivery to developing nations, WorldSpace's satellites are making use of traditional technologies in a novel combination. In general, there are two effective ways to allow broadcasts from multiple earth stations to share the capacity of a single satellite transponder (transmitter-receiver). The first is frequency-division multiplexing

For some reason, the student has been relieved of the responsibility of being a student. It is commonly feared that high school education has degraded, and it seems to be slipping into the colleges at the undergraduate level."

All in all, it is clear to Campanella that "there must be some rethinking about the quality of education in U.S. universities. I cannot find in the U.S. anything the equal of what I found in Germany at the University of Erlangen and at the Fraunhofer Institute, where many students go for honing their skills before entering German industry. I get the feeling that European universities now outstrip American universities in technical education. From my experience, especially in France and Germany, their technical skills are well-honed. In Europe, of course, there is a strong government hand in technology support, but it seems to allow latitude for accommodating freedom of thinking. Whether government support is good or bad, I won't say, but it does result in a lot of bright students in special institutes who move on into the national industry.

"I feel my university experience was good not only because I was a serious student, but also because I was lucky enough to have mostly inspiring, competent professors as teachers. Many universities today have good professors on their faculty, but rarely do they teach undergraduates. Instead, undergrad classes are taught by assistants who are inexperienced, not well versed in the subject, and not good communicators. Things would improve if universities would use only fully qualified teachers able to communicate—because it's in the undergraduate years that a student needs to get a good start."

◆

S. Joseph Campanella, who received his IEEE Fellow award in 1978 "for contributions to signal processing and satellite communications," holds the 1990 IEEE Behn Award in International Communications and has been inducted into the Engineering School Hall of Fame of the University of Maryland.

(FDM), in which the transponder's bandwidth is divided into different frequency bands, each of which is assigned to an earth terminal for its programming. The second is time-division multiplexing (TDM), in which the transponder's single carrier is divided into a repeating series of time slots or frames, into which each earth terminal places a burst of programming.

With the WorldSpace satellites, "the whole idea is to have an onboard processor with FDM channels coming up—which is the most energy efficient way to operate the uplink from an earth station, and then convert it to TDM—which is the most efficient way to handle the satellite power on the downlink," Campanella explained. The combination of the two techniques is the "ideal way to operate the satellite" because it is possible to "quadruple the power to the signal out of the satellite with the same solar power from the sun" while also having "high efficiency on the ground." It thus "gives high flexibility for developing countries to have small, low-power, economical earth stations uplinked to the satellites," Campanella said.

With the 1998 launch of WorldSpace's *Afristar* satellite, modern digital radio services were made available for the first time to developing nations in Africa with up to 100 satellite-direct digital broadcast carriers. [For Campanella's concerns—related to SDARS—about the future of U.S. higher education in the twenty-first century, see his box "What is the future of the U.S.'s universities and corporate research laboratories?" p. 136.]

Cyberwisdom

"There is a potential for larger intelligence forming on the 'Net," declared Robert W. Lucky of Telcordia, "a collective wisdom that has not yet been tapped. The 'Net is the world's largest library—with not only books but also people. If we can gather their collective wisdom on individual problems, fantastic things can happen."

As a mundane example, he cited national football pools, in which millions of people bet on the outcomes of football games. What fascinates Lucky is the uncanny accuracy of the national point spread—that is, the projected number of points by which a specific winning team will beat a specific losing team. "Somehow the collective wisdom *knows* how the games will turn out, even though no individual knows." To him, a football pool is "a simple numerical example of how the 'Net can put together everyone's imperfect information and come up with a complete picture that is accurate."

In his view, that phenomenon of "group filtering"—which has been studied at Telcordia and elsewhere—has remarkable potential for scientific and technological inquiry. "When you want information, often it's not 'book learning' you're seeking, but something from people's experiences." He cited the example of people who have just been diagnosed with cancer or some other serious illness seeking information and counsel on the Internet from others with similar experiences. But the collective wisdom and experience could accelerate the pace of science and engineering. "The state of the art is not in books and journals," he observed, in part due to publication delays. "It floats in the collective minds of the leading experts at the time." In his view, the invention of FAQs (frequently asked questions) is an ingenious mechanism for maintaining currency in a field. "Here are

As more advanced technologies are plugged or built into TV sets, and the Internet expands, the separate communications worlds encroach on each others' territories. Above is the main page for Worldgate Communications' Internet TV Over Cable service. At right is a Bloomberg financial news broadcast carried via Apple's Quicktime 4. For now, web access via cable is limited by the low resolution of TV screens, and TV via Internet is limited by bandwidth. *PRNewsFoto and Worldgate Communications (above) and Apple Computer (right)*

current answers as determined by experts, which change from week to week with the changing consensus view of experts at work," he said.

Cyberconsensus will also shape the role of the Internet in commerce, Lucky asserted. He is confident that electronic commerce and cybercash will become major forces in the twenty-first century, despite the negative feedback that delayed early attempts at their introduction in the late 1990s.

"Money is fundamentally a belief system," Lucky observed. "Things are worth what you *think* they're worth. We [in the United States] trust a dollar bill because of what we *believe* is behind it"—to wit, the full faith and credit of the Federal government. But every time money has been faced with a transition from one belief

ROBERT W. LUCKY: Why can't we better predict which technologies will succeed?

"We can definitely predict fantastic technology—but predicting what will be *done* with it goes off into socioeconomic chaos," declared Robert W. Lucky, corporate vice president of Telcordia (formerly Bellcore Applied Research) in Red Bank, New Jersey.

"What is difficult to predict is the future behaviors of markets, because that's fundamentally an instance of chaos in group dynamics," Lucky continued. "Metcalfe's law [named after Robert Metcalfe, inventor of the Ethernet local area network] is based on the observation that the value of a network grows as the square of the number of users—that every new user brings significant additional value to everyone else on the network. The overall value of the network to an individual user is small if few other users share it, but it becomes very large if many users are connected. Thus, in the adoption of a new network or service, there are two discernible regimes—a sparse regime in which there is little value to anyone and little incentive for new users to join, and a dense regime where there is great value for everyone and large incentive for new users.

"The huge barrier for any new service is getting from the first regime to the sec-

system to another, there has been a period of psychological shakiness until the new belief system was by and large accepted. U.S. examples include the historical shift from the gold standard to the silver standard around the turn of the twentieth century, and the shift from the silver standard to the Federal Reserve Note in the 1960s.

A transition from cold cash to electronic cash is no different, Lucky noted. In his view, "electronic money is in every sense much better than coins and bills: it cannot be forged, it's easily transferred, and you can put constraints on its use." But acceptance of cybercash fundamentally means "do you *believe* that bits on your

ond. It is a barrier that few services have been able to cross. In a paper of 1998, I wrote about how the sparse regime of Metcalfe's law stops things from happening—that no one wants to be the first person to buy a new communications application, because the initial value is so small that no one has sufficient incentive to purchase.

"But I missed what I now consider the most important point about Metcalfe's law. Lately, I've been more focused on the dense regime—how after a technology passes the start-up phase, there is very strong positive feedback and great incentive for new users to join. It's a lock-in phenomenon where the winner takes all and the losers get nothing. We've seen it many times—the Windows PC versus Apple, VHS versus Betamax, Microsoft Word versus WordPerfect and other word processing programs, and Powerpoint versus other slide-making software.

"What's key is: *it does not matter which technology is better*—you have to go with the group," Lucky stated emphatically. "The whole group now wants to do packet communications, Windows, TCP/IP. Everyone has incentive to do what everyone else is doing. Intel, Microsoft, and the Internet have benefited from the lock-in phenomenon of Metcalfe's law.

"Economists say this phenomenon is descriptive of 'tippy' markets—once the markets are tipped one way, they are gone forever. Moreover, there is no *a priori* way to determine what will tip a market. It's a fundamental instance of chaos in group dynamics. And that makes it fundamentally difficult to

Who needs an HDTV set in the house when you can wear it? For as little as $900, Sony is selling Glasstron, an audio-video headset that replicates the viewing experience of a widescreen (1.3 meter) TV at 2 meters. Sony executives Dan Nicholson and Federico Stubbe show off the portable 150-gram headset that connects to VHS players, digital video, or computers. The image is produced by 18-millimeter, 180,000-pixel (video) to 1.55-million pixel (computer SVGA) liquid crystal displays with special lenses to project the image. *Sony*

predict future societal behaviors in the adoption of technologies.

"Even in the monopoly era in telecommunications when the Bell System could semi-determine its destiny without great regard for markets, Metcalfe's law was at work. Beginning in the 1960s, the telecommunications industry had three eras of visions—videotelephony, home information systems, and video on demand. All of them were faulty—and each one drove the design and philosophy of the network for the next 10 years. That seems almost ludicrous today. Today, nobody plans because nobody is in control. In a free market, you risk that if a technology works, you get your money back—and if it doesn't, you lose. The era is past when you can force people to pay for *your* future, whether they want it or not.

"Metcalfe's law guarantees that we'll never be able to predict which technologies will become market successes."

> ' . . . chaos in group dynamics . . . makes it fundamentally difficult to predict . . . the adoption of technologies.'

◆

Bob Lucky, who received his IEEE Fellow award in 1972 "for contributions to the theory and practice of data communications," invented the adaptive equalizer, a key enabler of modern data modems. He writes the popular "Reflections" column in IEEE Spectrum *magazine—from which many of his essays were taken and compiled in his book* Lucky Strikes Again *(Piscataway, N.J.: IEEE Press, 1993).*

hard drive are redeemable for purchases?" When a large number of people come to believe that, he feels, then electronic money will become prevalent. "It all goes back to Metcalfe's law and the effectiveness of capturing markets," he added [the subject Lucky explores in detail in his box "Why can't we better predict which technologies will succeed?" on page 140].

Predictions: Action at a Distance

Lucky also foresees the possibility of "sending physical objects over the communications network"—that the Internet will eventually blur the distinction be-

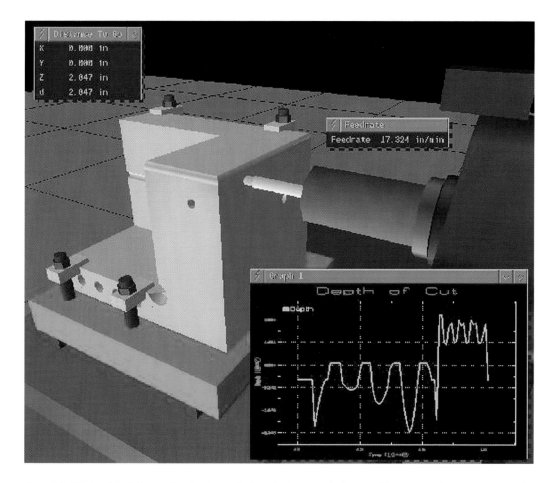

Deneb's Virtual NC is a physics-based simulation tool that utilizes actual computer code so operators can check clearances and processes before committing precious raw stock to the factory floor. Such technology can also be used to command action at any distance, in the next room or the next continent. *PR NewsFoto*

tween bits and atoms. "A fax machine is a crude example," he said, in allowing a copy of an original document to be created at a distant location. He sees the day, not too distant, when it would be possible to "send bits to direct a machine tool at the other end of the line to build a three-dimensional object, such as an electric motor, rather than to manufacture it and truck it halfway around the world."

The last frontier, of course, is the transmission of gene sequences for the creation of life at a distance. "Ultimately, it's information that creates a living being," Lucky observed. "DNA is a blueprint. You and I are just a series of bits, instantiated." If he is right, then twenty-first-century life would imitate twentieth-century art: half a century ago, just such a hypothesis was the premise of British astronomer Sir Fred Hoyle's 1962 novel *A for Andromeda*, in which radio signals received from the Andromeda galaxy were found to be biological code that allowed the *in vitro* creation of a woman (a concept revised in the 1996 horror thriller movie, *Species*). ◇

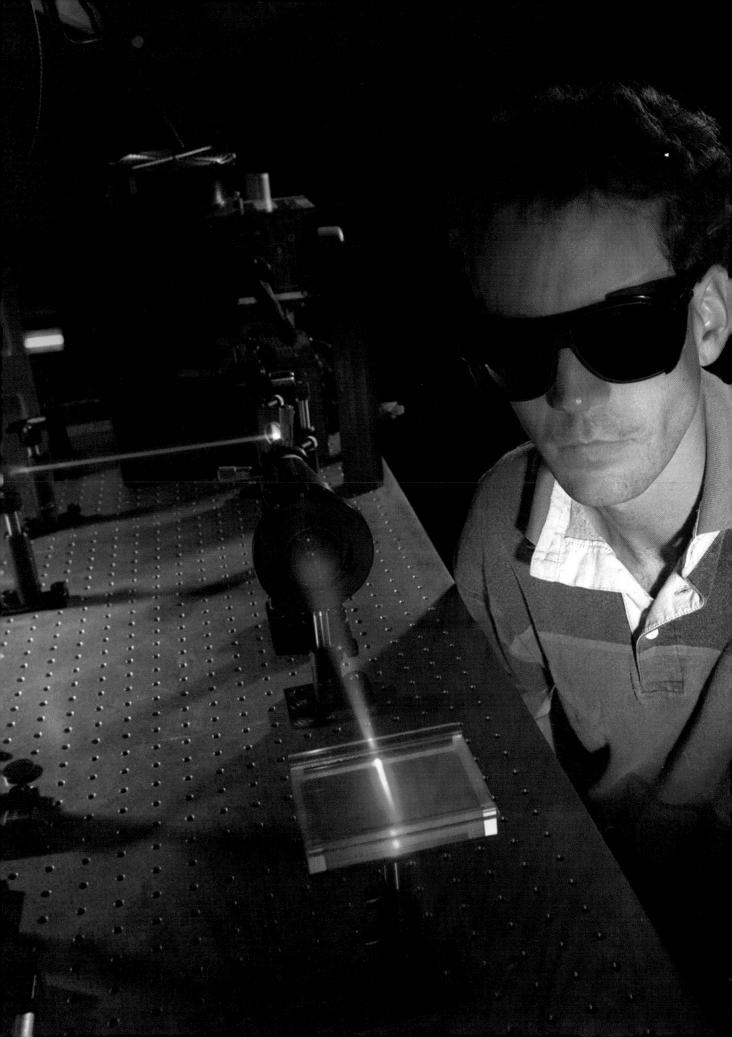

CHAPTER 7

MEDICINE AND BIOLOGY

CURRENT MEDICAL AND BIOLOGICAL technology already reads like science fiction. That projected for the twenty-first century sounds even more fantastic. If all comes to pass as serious biomedical investigators and engineers envision, age-old handicaps that used to devastate people's lives—blindness, deafness, paraplegia, memory loss, and even death—may be circumvented by intelligent prostheses, neural implants, artificial organs, natural organs regrown after injury, and possibly even suspended animation. Moreover, exploratory surgery has been almost completely replaced by advanced medical imaging.

Here, with perspectives on twenty-first-century possibilities for biomedical engineering, are four experts in very different fields. Thelma Estrin, professor-in-residence emerita in the computer science department at the University of California at Los Angeles, was an early pioneer (1950s and '60s) in the use of digital techniques for medical monitoring. Cato T. Laurencin, Helen I. Moorehead Professor of Chemical Engineering at Drexel University and director of the university's Center for Advanced Biomaterials and Tissue Engineering in Philadelphia, Pennsylvania, is at the forefront of engineering regrowth in bone, cartilage, and liga-

Left: Oak Ridge National Laboratory chemist Eric Wachter is part of a team experimenting with the two-photon laser and its medical applications. The new laser technique allows for precise targeting of tissue. The laser doesn't damage skin or adjoining tissue. Also, unlike X-rays, the two-photon laser is intrinsically safe. Wachter and colleagues in different disciplines at ORNL believe their two-photon laser technique has great potential for treating breast cancer. *ORNL Photo by Tom Cerniglio*

ments. Ray Kurzweil, CEO of Kurzweil Technologies Inc. in Wellesley, Massachusetts, and founder of four companies that bear his name, is an inventor of a number of ground-breaking artificial intelligence technologies since the 1970s, including reading machines, music synthesizers with realistic sounds, optical character recognizers, and speech recognizers. And George S. Moschytz, director of the Institute for Signal and Information Processing at the Swiss Federal Institute of Technology in Zurich, Switzerland, has thoughts on various techniques for helping the profoundly deaf to hear.

Growing New Organs

"Tissue engineering is amazing," exclaimed UCLA's Thelma Estrin, who is excited about the products on the market (including engineered skin for burn victims and for the ulcers of diabetic patients) and the products in the making (which may shortly include pancreas and liver).

"Tissue engineering combines the principles of biology, chemistry, and engineering to repair, restore, and regenerate organ tissues," explained Drexel University's Cato T. Laurencin, who is not only a medical school professor, but also a pioneering researcher in the tissue engineering of musculoskeletal tissues and a practicing orthopedic surgeon. Modern tissue engineering started in the 1960s and 1970s with the search for synthetic materials that could substitute for natural tissue. Although synthetics do work, "they wear out," Laurencin said. "So it's better to design a synthetic structure that the body recognizes as its own and can slowly remodel and replace with its own structure." The driving forces behind the investigation of synthetic tissues are the short supply of natural tissues from live donors or cadavers, the demand for tissues by patients the world over, and physicians' growing apprehensions about disease transfer through slowly replicating viruses that might be lurking in natural biological tissues.

Laurencin's work has primarily been devoted to studying the regrowth of bone, cartilage, and other tissues of the musculoskeletal system. "Bone is important. There are literally millions of bone graft operations performed worldwide every year" to replace bone lost in fractures, in total joint replacements (especially hips and knees weakened by age), and cranial/facial replacements.

"The current gold standard is the use of autograft tissue—that is, tissue from the patient's own body," Laurencin explained. "But the body has a limited supply of bone, and the operation to remove some—typically from the iliac crest above the hip bone—is very painful. So if you wish to perform a bone graft onto, for example, the patient's humerus [upper arm bone], three months later the patient is probably going to be complaining about pain in the hip area from where the

NASA's rotating wall vessel, or Bioreactor, literally puts a new spin on cell growth by rotating to suspend cells in an environment mimicking weightlessness for several weeks until the assemblies grow too large to remain suspended. At that point, experiments are continued in space where the cells are weightless and the rotation

ensures cell nourishment. Ground tests have yielded samples that resemble natural cell clusters rather than the pancakes normal to cultures. In an experiment flown

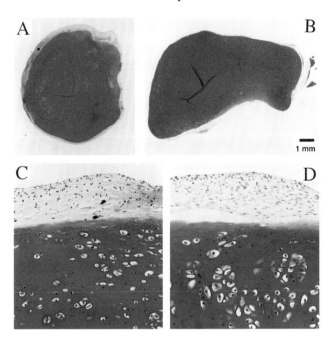

Sept. 16, 1996–Jan. 22, 1997, investigators at the Massachusetts Institute of Technology reported that cells differentiate and grow into distinct structures, demonstrating the feasibility of tissue engineering in space. Images A and C show a 10-power and a 200-power views of samples grown in space; B and D show samples grown on Earth. *NASA/Johnson Space Center (above), MIT and Proceedings of the National Academy of Sciences (left)*

bone graft is taken." Moreover, since 30 percent of such patients are also plagued with infections and other complications, there is "a real need for an alternative."

Laurencin's lab—first at MIT and then at Drexel—was one of the first to design synthetic bone that could replace natural bone that is lost. Its ultimate goal is to develop systems for regenerating whole limbs, involving complex muscular and skeletal tissues. Laurencin, along with two colleagues (Paul Lucles and Henry Young), is investigating muscle morphogenetic proteins, which "when placed subcutaneously, can start to form muscle," Laurencin said. He is also conducting studies examining bone morphogenetic proteins that similarly start to form bone. "So we have the elements for the beginning of the regeneration of skeletal tissues," he said.

"The Holy Grail may be cartilage," he continued. "There are a wide range of

CATO T. LAURENCIN: How can mentoring overcome racial discrimination?

"Everyone would like to believe that racial discrimination is of the past. But it still goes on now at every level of life," observed Cato T. Laurencin, Helen I. Moorehead Professor of Chemical Engineering at Drexel University in Philadelphia, Pennsylvania. "I've felt it, yes, as a student [in the 1970s and '80s] when being stopped and questioned when I was heading across campus to do my studies. And even as a surgeon, once I was handed a mop when coming out of the operating room after performing a surgery.

"But I'm someone who tries to be positive. I don't dwell on it. This [the United States] is a great country, a great environment for change," he continued. "But I also don't

types of cartilage defects that need to be treated." They range from trauma in motor vehicle accidents (especially when a person's knees impact the dashboard) to chronic diseases such as osteoarthritis (the most common form of arthritis, in which cartilage is lost due to biochemical and biomechanical changes in joints).

The knee, one of the largest joints in the body, is the focus of much research in tissue engineering in cartilage. "Cartilage normally grows very slowly, with not a large amount of supporting matrix," Laurencin said. But now, various labs are experimenting with harvesting a patient's cartilage, growing it more quickly *in vitro*, and transplanting it at some later time.

discount discrimination, either." Laurencin's efforts to combat discrimination are to encourage fellow African Americans and other minorities to enter science, technology, and medicine.

"I am dedicated to mentoring," he stated simply. His laboratory makes a special point of training minority engineering students interested in biomedical engineering, medical students interested in orthopedic surgery, and science students interested in research

'I've felt discrimination, yes—once I was handed a mop when coming out of the operating room after performing a surgery.'

work. Beyond the graduate and undergraduate students at Drexel, Laurencin also takes as many opportunities as possible to speak with students in high schools, junior highs, and even elementary schools on the potential of technical careers. He is team physician for inner-city schools and also a physician for USA Boxing, the certification organization for amateur boxing, since boxing is a popular sport in many inner-city gymnasiums. ("It's an art form with great mental discipline; boxers are taught to respect their bodies and themselves, and must adhere to strict practice schedules with no drugs, no alcohol," Laurencin stated.)

From his own personal experience, he knows the crucial role mentoring can have on a young person's life and opportunities. "I've modeled my life after two people," he recounted. "The first is my mother, Helen I. Moorehead [1922–94], who was a primary care doctor in Philadelphia for almost 50 years. She was not only a fine, caring clinician who delivered babies and took care of the aged; she was also a scientist with a small laboratory in the back of her office. And she was a great mentor to young physicians in

Among other things, tissue engineering is also focused on developing tools of use to paraplegics, including synthetic ligaments and the development of extended graft systems for nerves. The effort to marry neurons and microelectonics into hybrid circuits is still in its infancy. "There are great challenges remaining in the tissue engineering of musculoskeletal tissues and the central nervous system," Laurencin said. "The next century will be all about bringing medical and engineering technologies together—and bringing the results to patients." [For more about the special role of Laurencin's lab in mentoring the careers of minority physicians and biomedical engineers, see his box "How can mentoring overcome ra-

'We must encourage more formalized mentoring programs to encourage African Americans and other minorities to enter careers in science, and to be able to succeed at a very early age.'

the city, providing many scholarships to students." Like her, Laurencin is now a clinician (an orthopedic surgeon specializing in shoulder surgery and sports-related injuries), a research scientist (specializing in tissue engineering), and a professor.

"The second person is my scientific mentor, Robert S. Langer, who is now the Kenneth Germehausen Professor of Chemical Engineering at the Massachusetts Institute of Technology," Laurencin continued. "At the time I was doing my Ph.D. under him, he was young and not very well known. But now he is recognized as the father of tissue engineering and one of the fathers of polymer drug delivery. Yet, even though he's one of the greatest scientists of the twentieth century, he's stayed modest and humble. He has probably trained more scientists in these areas than anyone else in the world," he observed.

From both seeing and experiencing the influence of his mother and of Langer on himself and other young people, Laurencin is convinced that "mentoring is vitally important." He feels it is especially important for African Americans and other under-represented groups who, even on the cusp of the twenty-first century, still have comparatively few role models and supporters. "We must encourage more formalized mentoring pro-

cial discrimination?" p. 148. And see Thelma Estrin's thematically related box "What practical advice can encourage women engineers?" p. 154.]

Restoring Control to Paraplegics

Marriages of computer hardware and biology have taken many forms: from the devices that allowed Estrin to study the electrical activity of the brain, and Wilson Greatbatch to study—and then mimic—the electrical activity of the heart, to the interests of the Swiss Federal Institute's George S. Moschytz. His interest is electromyography—the interaction of muscles and electronics.

grams to encourage African Americans and other minorities to enter careers in science, and to be able to succeed at a very early age."

He also feels it is important for minorities to gain a deep appreciation for the history of science, technology, and medicine. "Africans and African Americans have an incredibly rich history in science around the world," he said. "But too often, it's not been promulgated." As just one example related to medicine, he cited the early twentieth-century African American chemist Percy Julian, who started his own company and developed a technique for synthesizing cortisone—a technique that has since become the basis for synthesizing a variety of other steroids.

"In the next century, the haves and have-nots will be decided more by technological competence than by money or family background," Laurencin concluded. "Computers and the Internet are becoming a staple of society. We must make sure that every young person, regardless of race, can be part of the haves. The racial discrimination problems we're facing now as a nation will either be solved in the next century—or will come to a head."

◆

Cato T. Laurencin has a B.S.E. in chemical engineering from Princeton University (1980), an M.D. (magna cum laude) from Harvard Medical School (1987), and a Ph.D. in biochemical engineering/biotechnology from the Massachusetts Institute of Technology (1987). In 1995, he was the first physician to receive a Presidential Faculty Fellow Award from President William J. Clinton. In 1998, he was one of five Americans to receive the American Orthopaedic Association's prestigious ABC Traveling Fellowship, to give six weeks of lectures internationally at academic centers throughout the English-speaking world. He is the author of 150 presented and published papers, and the holder of a dozen patents.

"Muscles are activated by the nerves through signals from the brain," he said. "If something is wrong with that connection—with the nerve itself or its shielding, as happens with various neuromuscular diseases [such as multiple sclerosis and cerebral palsy, and with injuries to the brain, spinal cord, and peripheral nerves]—then the person becomes palsied or immobilized. The firing rate and the form of the signal are crucial."

So for some years, Moschytz has been analyzing signals from muscles and using signal-processing algorithms and sophisticated techniques to isolate specific signal components. "It's been very helpful for diagnostics, as such analysis gives

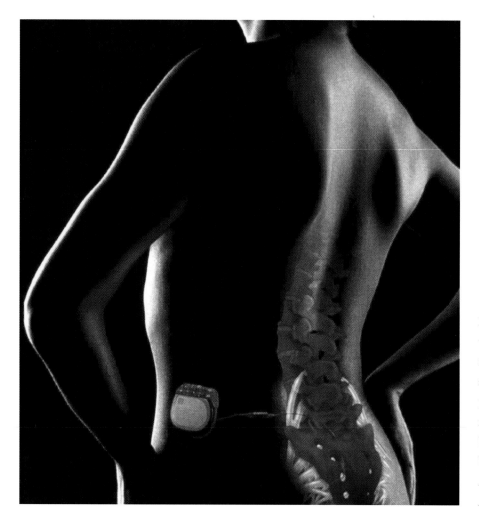

InterStim therapy for urinary control uses neurostimulation to send mild electrical pulses to the sacral nerves in the lower back that control bladder function. *PR NewsFoto and Medtronics*

indications of neuromuscular activity and—if defective—can isolate whether the problem lies with the muscle, the nervous system, or the brain." Ultimately, he feels that a better understanding of the neuromuscular system, aided by improved methods of electromyography, will help paraplegics walk. Already, signals derived from the upper body are being used to stimulate muscles in the lower body "so the person can stand up and—with the aid of supports—walk several hundred meters," he noted. A recently commercialized device uses just this sort of upper body information and appropriate signal processing to allow quadriplegic patients control of one of their hands. Further, the U.S. Food and Drug Administration has recently approved implants that provide electrical stimulation to the urinary bladder to allow paralyzed people to have improved bladder control. [For Moschytz's views on how engineering can help humankind, see his box "How can we ensure that technology is humane and not inane?" p. 160].

Meanwhile, Kurzweil Technologies' Ray Kurzweil is taking a completely different approach in helping paraplegics take their first steps. He is working on a design for an "exoskeleton orthotic device that could allow paraplegics to walk

These ultrasound images show the chambers of the heart before (top) and after (bottom) the injection of a contrast agent that allows doctors to see more of the fine vascular structure of the heart—or more blockage where expected structures do not appear. With Power Harmonics ultrasound technology, the contrast agent can be visualized in the tiny blood vessels of the heart wall, an indicator of healthy heart tissue. *PR NewsFoto and Advanced Technology Labs*

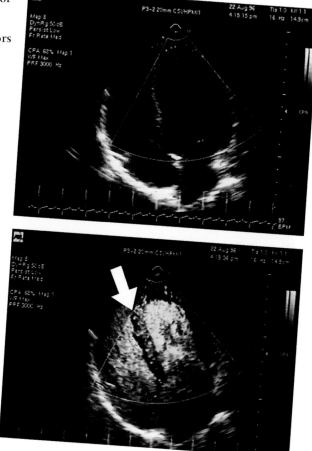

with a fairly normal gait, navigate steps, and overcome the limits of wheelchairs," he said. All it would require from its user is "some flexibility in the joints."

Interestingly, his device—which is still on the drawing board—does not rely on electrical stimulation of the person's own leg muscles or nerves, so it "could also work with completely artificial limbs." Instead, Kurzweil said, it is intended to be a "completely noninvasive compact device that a person "could put on as easily as clothing." The prosthesis would provide external mechanical movement, while the legs themselves would provide passive mechanical support. The person would direct the device's movements by controls on the grip of a specialized cane.

Restoring Sight and Sound

UCLA's Estrin sees tremendous promise in the twenty-first century for implantable devices, especially for handicapped people. "Cochlear implants are allowing the profoundly deaf to hear," she pointed out, "while eyeglasses for the blind will pick up light and send signals to the brain."

One such area is the restoration of sound to the profoundly deaf—those whose cochlea (a spiral structure in the inner ear) has deteriorated either through disease or assault (such as happens in teenagers exposed to too much 100-decibel music). Today's cochlear implants are "still very crude," said Moschytz, primarily because the cochlea is so complex—"it is actually a more complicated organ than the eye

THELMA ESTRIN:
What practical advice can encourage women engineers?

"Some women still feel oppressed or unequal in an engineering environment," observed Thelma Estrin, professor-in-residence emerita in the computer science department at the University of California at Los Angeles. "If they had a little more insight into sociotechnology, they might realize it takes decades to change social systems."

Estrin herself had to face the male prejudices of a largely patriarchal engineering profession when she received her Ph.D. from the University of Wisconsin at Madison in 1951. But she not only pioneered the use of digital sampling and processing techniques in the then-analog world of electroencephalography (study of the electrical functioning of the brain) in the 1950s—she also made engineering a personal family affair. In the late 1940s, she married fellow engineering student Gerald Estrin, whom she subsequently

because it does so much signal processing before passing the auditory signal on to the brain." Cochlear implants consist of electrodes that are surgically placed near auditory nerves in the inner ear and that respond to different frequencies. A tiny microphone outside the ear picks up sound and transmits it to a signal processor placed behind the ear. After some analysis, the processor sends the processed signal to an implanted electrical generator that then stimulates the appropriate combination of electrodes that, in turn, stimulate the auditory nerves. Cochlear implants are by no means perfect—but profoundly deaf people with such prostheses can typically carry on a conversation.

helped to build Israel's first computer (the WEIZIAC, for the Weizmann Automatic Computer at the Weizmann Institute of Science in Rehovot). Two of their three daughters went on to achieve independent renown in computer science (Judith as chief technical officer of Cisco Systems and Deborah as professor of computer science at the University of Southern California); the third (Margo) is a physician in private practice, specializing in internal medicine.

As a result of her own personal experience and perspective, Estrin is keenly interested in helping universities encourage women in science and engineering to become aware of women's studies programs and the history of women in professions.

'You cannot instantly change the attitude or behavior of many individual men who've grown up in a different era or society. You have to laugh at some of it and report the rest.'

Estrin's advice to young women contemplating a potential career in engineering is solidly down-to-earth. "First, most engineering jobs are helpful to society, give the opportunity to create something new, and are interesting and rewarding with good chances for advancement. Second, engineering jobs are everywhere—which, society being what it is, helps women who are looking for jobs where their husbands have to go. Third, I am not of the school of thought that a woman—or any other engineer—needs to have an A in math or physics. But you do need to understand the subjects enough to use them practically in the wide variety of options and employment available in technology."

Reflecting on how far the engineering profession has progressed in "gender enlightenment" just in her own lifetime (she was born in 1924), she cautions young women to

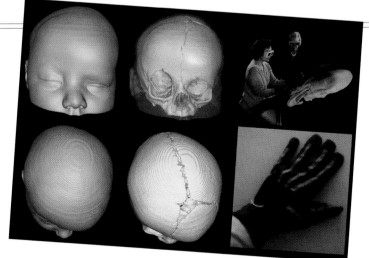

Virtual reality modeling will allow surgeons to explore injuries and map reconstruction plans before operating on a patient. In a joint project between NASA's Ames Research Center and Stanford University, a girl born with a deformed skull was examined by a conventional CT scan, which then was assembled into a VR model showing both soft tissue and bone. Researchers then explored the skull using a virtual workspace desk and a "pinch" glove that provides tactile feedback to reinforce the images. *NASA/Ames Research Center*

'You need to *like* your field and be excited by what you're doing—and hope you are contributing to better society.'

be less impatient about the lack of complete societal transformation. "I feel the whole male/female environment in work and society is indeed changing," she said thoughtfully. "But it will take another generation or two for *all* attitudes to change" to that of complete equality. "You cannot instantly change the attitude or behavior of individual men who've grown up in a different era or society. You have to laugh at some of it and report the rest to proper authorities if their actions are very disturbing or illegal."

Most important, she feels, "you need to *like* your field and be excited by what you're doing—and hope you are contributing to better society."

◆

Thelma Estrin, who received her IEEE Fellow award in 1977 "for contributions to the design and application of computer systems for neurophysiological and brain research," was the sixth woman to attain this honor. She was director of the Digital Processing Brain Research Institute of the University of California at Los Angeles from 1970 to 1980. She was also the first woman to be appointed to the Board of Trustees of the Aerospace Corp. in Los Angeles, in 1978. She was president of IEEE's Engineering in Medicine and Biology Society (1977), and was the first woman to be elected to the IEEE's board of directors (1979), on which she served for several years.

One problem with cochlear implants, however, is that the signals reaching the inner-ear auditory nerve endings are still very different from those that would be transmitted to the brain in a hearing individual with a healthy inner ear. Moreover, cochlear implants require invasive surgery, with some risk of complications.

Moschytz's interest is not so much directed at improving cochlear implants as to exploring noninvasive pathways for getting auditory stimuli to the brain—specifically, by tactile stimulation of the hand or forearm. "The brain has been mapped, and certain areas respond to visual and auditory stimuli," he explained. "Fifty years ago, it was found that the nerve endings on the forearm have a pretty large area

Tuan Vo-Dinh's "DNA biochip" could revolutionize the way the medical profession performs tests on blood. The chip, for which a patent has been filed, will provide for instant test results for the AIDS virus, cancer, tuberculosis and other diseases. Vo-Dinh, a member of Oak Ridge National Laboratory's Life Sciences Division, expects the chip to also have environmental applications. *ORNL photo by Tom Cerniglio*

reserved for them. So if you stimulate the surface of the forearm skin acoustically, either through mechanical vibration or through tiny electric shocks, you can activate the part of the brain cortex reserved for tactile stimulation of the forearm, and 'teach' the deaf person to translate this stimulation into acoustical information."

Moschytz and his colleagues have designed some devices that give the rhythm of speech or music on the arm, which could, for example, be useful to deaf mothers who could not hear when their infants started crying in their cribs. Vibrotactile devices have been shown to be helpful in assisting profoundly deaf people in reading lips, or even in being the main conveyor of acoustic information.

Similar noninvasive aids for people with lost vision also would be invaluable. According to Kurzweil, "reading machines that are pocket-sized" would be useful tools. Such portable devices could snap pictures of text and instantly read it aloud, allowing a blind person to "read street signs, soup cans in grocery store aisles, LED displays, computer screens"—in short, all the text that sighted folk routinely encounter during the day.

The next step after that would be "seeing machines that you direct by moving your head, that describe the world to you, like an intelligent friend whispering in your ear," Kurzweil went on. If linked to the transmissions from the constellation of Global Positioning System (GPS) satellites in orbit around the earth, such a seeing machine would "help you navigate and avoid obstacles." Initial versions of such a seeing machine could indeed be mounted on standard eyeglasses.

But the desideratum, of course, would be neural implants "that create a virtual visual sense, that literally overcome blindness," Kurzweil said. His 1999 book, *The Age of Spiritual Machines*, describes experiments with the design of retinal implants, for example, that would "not only replicate the retina but also do early visual processing such as that done by the optic nerve and the brain. Thus, the retinal implant could create sight even for people whose visual cortex had been destroyed by disease, stroke, or injury, or atrophied through lack of use." Implants that electrically stimulate the retina and allow people to perceive shapes and letters are just now being reported in the literature. Ultimately—within several decades, projected Kurzweil—there may be "all kinds of neural implants that could extend the quality of all the senses, to overcome the natural deterioration of aging—and to extend the normal range of sensory experience," he said. [For more of Kurzweil's predictions of what the twenty-first century might hold for humans and machines, see his box "When computer intelligence exceeds human intelligence, what will it mean to be human?" p. 166.]

What You See is What You Get

Advances in medical imagery are rendering "exploratory surgery" a procedure of the past. Even more advances are projected for the twenty-first century. Just one such advance, now in laboratory prototype, is optical coherence tomography.

Using light and mirrors, researchers have developed a minimally invasive way to peer inside the blood vessels of the heart. No better images could be obtained if the sections were surgically removed and placed under a microscope. Relying on fiber optic cables to steer infrared light, investigators direct a catheter through the veins into the heart to the region of interest. Infrared, whose wavelengths are longer than visible light, can penetrate several millimeters into tissues that are opaque to visible light and generate images with a resolution of 4 micrometers—less than a tenth the width of a human hair and fine enough to reveal the structures of individual cells. The only incision is the nick that allows entry to the blood vessel.

When commercially available, it is expected that such an "optical biopsy" will have 10 times the resolution of magnetic resonance imaging (MRI) or high-frequency ultrasound imaging—and may become the tool of choice to study the brain, the coronary arteries, and maybe even disorders of the central nervous system.

Targeted Drug Delivery

Molecular nanotechnology—a proposed manufacturing technology that should allow the fabrication of most arrangements of atoms that are consistent with physical laws—could have enormous implications for both tissue engineering and drug

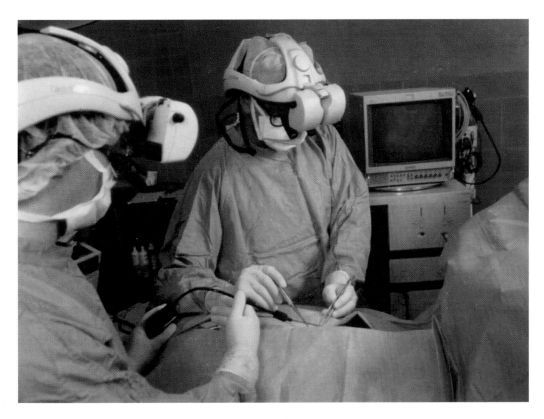

These surgeons perform minimally invasive surgery while wearing Vista's CardioView head-mounted display designed specifically for cardiothoracic surgical use. It provides 3D visualization of an endoscopic or microscopic image, and presents images to a surgeon in an optically correct, intuitive and ergonomic fashion. *PR NewsFoto and Vista*

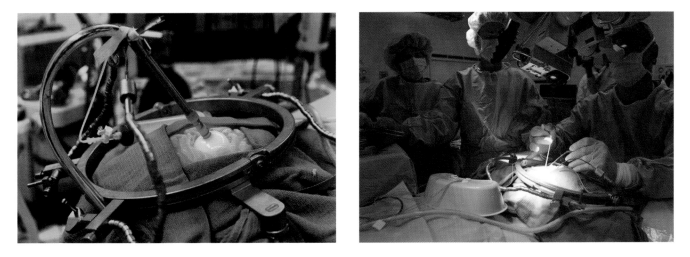

What started out as an attempt to develop a light which would allow for the growth of plants in space led to a remarkable tool used by surgeons in the fight against brain cancer in children. In the system being developed by NASA, the patient is injected with Photofrin II, which attaches to unwanted tissues and permeates into them. Next, a tiny balloon is inserted into the cancerous area and inflated so the tumor cells can be illuminated by light emitting diodes shining through optical fibers. This activates the drug which kills the tumor cells and leaves healthy tissue unaffected. *NASA/Marshall Space Flight Center*

GEORGE S. MOSCHYTZ: How can we ensure that technology is humane and not inane?

"One need only mention the names of disasters such as Chernobyl, Bhopal, Seveso, the Exxon *Valdez* to lead us, all too easily, to conclude that technology is at the root of all our troubles," observed George S. Moschytz, director of the Institute for Signal and Information Processing at the Swiss Federal Institute of Technology in Zurich, Switzerland. These four major technological/environmental calamities of the 1970s and '80s spread nuclear or chemical pollution over huge regions, poisoning people and animals and rendering the land temporarily or permanently uninhabitable. "But I believe that technology has a key role in alleviating critical human problems—including those created by technology itself.

"Most technology originated to cater to basic human needs. Such humane technology includes sanitation, electric lighting, medical instrumentation, and mass transportation. Moreover, the inventions of the electric motor and the steam engine alleviated human labor and drudgery and, most important, helped abolish slave and child labor.

delivery. That is the belief of nanotechnology pioneer Ralph C. Merkle, research scientist at Xerox Palo Alto Research Center in California. [For some technical background on molecular nanotechnology, see Chapter 2: Structures and Devices.]

"One intriguing idea is the use of nanotechnology to build very good artificial red blood cells," Merkle suggested. One of the primary purposes of red blood cells is to transport oxygen throughout the body. But certain illnesses and injuries, such as arteriosclerosis and stroke, impair or block cardiovascular circulation so that parts of the body—including the brain—may be deprived of normal amounts of oxygen. The amount of oxygen stored in the red blood cells' hemoglo-

"However, the indiscriminate and unchecked use of successful technological innovation can create its own problems," Moschytz warned. "This is hardly news—back in 1976, problems caused by technology were called the 'failures of success' by the late Herman Kahn [founder of the Hudson Institute in Indianapolis, Indiana]. For example, consider the clogged highways and air pollution created by the automobile.

"But usually, technology can solve its own self-generated problems—and may even solve the problems created by other technologies. For example, the recognition that a single telephone wire can accommodate far more than a single telephone conversation led to the search for methods of multiplexing—that is, of sending hundreds of messages over one transmission channel. Today, high-capacity optical-fiber communications qualifies as a humane technology that neither pollutes the environment, nor disfigures the cities or countryside. Conference calls from the office, or from homes in the suburbs, may eventually ease clogged highways by reducing the need for business travel or daily commuting," Moschytz said. "Moreover, through mass communications, the plight of refugees from a flood or an earthquake in any part of the world is immediately brought to the attention of millions—and has repeatedly triggered spontaneous, large-scale programs of humanitarian assistance from private citizens and governments thousands of miles away.

"Yet, indiscriminate technological innovation leads to *inane* technology. We should push a successful design as far as is desirable—but not necessarily as far as is *possible*," Moschytz said. "We need to accept the fact that just because something *can* be done, it does not necessarily *have* to be done.

"By which criteria should limits be set?" he asked. The standard is simple. "*When*

> 'We should push a successful design as far as is *desirable*—but not necessarily as far as is *possible*.'

bin molecules is "relatively low," Merkle noted. "You can't hold your breath very long, because your red blood cells don't carry much oxygen."

The artificial red blood cell Merkle envisions could be somewhat smaller than a red blood cell—only 1 micrometer across instead of 8 micrometers. Chemical engineers have long known of clathrate or "basket" molecules that can carry oxygen and release it when the oxygen partial pressure is low enough—but Merkle's concept is more ambitious. His artificial red blood cells would consist primarily of a spherical tank of diamond, "which would be strong enough to hold oxygen at 1,000 atmospheres," he said. "It would absorb oxygen from the lungs, compress it

'When technology is not primarily motivated by human needs, there is no legitimate market and demand must be generated artificially. Such technology is inane.'

technology is not primarily motivated by human needs, we will be blindly creating technology for which there is no legitimate market and for which a demand must be generated artificially. Such technology is *inane*: it ignores, is essentially irrelevant to, and is even contrary to, human needs. Inane technology does not solve existing human problems; on the contrary, it more often causes new ones by running rampant by following its own laws of growth and self-perpetuation. It has no worth for society at large, and causes problems that have neither excuse nor justification.

"Resulting public indignation is understandable. The public rightfully feels that the human gift of innovation and creativity as embodied in technology is being abused. The first to blow the whistle in order to halt development of inane technology should be engineers and scientists themselves; it is they, as originators of technology, who must primarily become sensitized to what constitutes humane or inane technology."

In Moschytz's opinion, there are two steps that, if followed universally by people responsible for technological development, would effectively improve the thrust of technology towards making the world better.

"First, we must make a *continuous critical and discriminating appraisal of our accomplishments: Are we creating humane or inane technology?* Whether a technology is humane or inane often depends on how it is applied," Moschytz said. "For example, it would be absurd to propose doing away with television: anyone who has had to deal with the aged, the infirm, or the bed-ridden knows what a blessing it is to be able to bring the outside world into the living room. Yet how much further effort should be put into ever more

into the tank, be carried through the circulatory system to the site, and trickle out its oxygen into the tissues at lower pressure." One liter of such artificial high-pressure cells would suffice to meet all the body's oxygen requirements for a day.

Such artificial red blood cells could also absorb and compress carbon dioxide from the tissues and release it in the lungs to be exhaled. "And if you exchange oxygen and carbon dioxide simultaneously, you can balance the energy requirements for compressing one by decompressing the other," Merkle noted. "You could hold your breath for a day! And you could maintain tissue integrity even though the circulatory system is compromised," he added. "And if your circulation stopped

elaborate three-dimensional, holographic, life-sized or mini-screen television sets? How much effort should be devoted into developing video packs of every conceivable description, a video telephone, and so on? These are questions that should *not* be left to the public relations experts who, by hook or by crook, will try to convince the public of their desirability.

"Second, we must *actively seek and identify basic human needs that are amenable to technological aid*. As long as 'making a fast buck' is the *dominant* motivation for innovation, companies will not have the inclination to look for such applications because they are not necessarily very lucrative. A reading aid for the blind or an alarm detector for the profoundly deaf will likely sell in far smaller quantities than a personal computer or a video game. Nonetheless, I believe that there are many more applications than is presently realized, that would be *sufficiently profitable* for industry to pursue, whilst leaving their beneficial mark on society as a whole."

In short, Moschytz concluded, "If engineers heed their consciences, they will find a multitude of urgent, unsolved problems amenable to technological aid—whose solutions will do credit to the human gift of creativity, to technology, and to themselves."

◆

George S. Moschytz, who received his IEEE Fellow award in 1978 "for contributions to the theory and the development of hybrid-integrated linear communication networks," has worked on projects ranging from the transmission of color television signals to the development of hearing aids for the profoundly deaf. He is the author of more than 250 papers and four books, and the holder of several patents, in the field of network theory, active and switched-capacitor filter and network design, and sensitivity theory.

completely, even for hours, the artificial red blood cells would provide oxygen to your tissues and remove carbon dioxide." From the artificial red blood cell, it's only a small step to imagine similar spheres that could transport other metabolites (such as sugars) or remove wastes (such as urea—thereby eliminating the need for external kidney dialysis), he suggested.

Another potential use is the targeted delivery of medications to kill individual cancer cells without injuring surrounding tissues, Merkle noted. One method for guiding nanostructures to the tumor site is to place an external ultrasound signaling device on the skin over the tumor for the spheres to home in on. Alter-

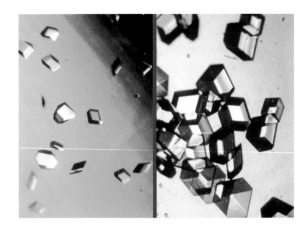

Crystallizing proteins in space, away from the disruptive effects of Earth's gravity, is allowing more detailed insights into the heavy, complex proteins that regulate much of life. X-ray crystallography of recombinant human insulin crystals grown on Earth (top left) left questions about the structure of the molecule as depicted in the wireframe model of electron clouds that outline the molecule (bottom left). Space-grown crystals (right, top and bottom), which were larger and more ordered yielded fine details showing one large molecule rather than two small ones. The improved understanding of the structure and links will allow production of insulin suspension that ensures a smoother absorption of the injected microcrystals. *NASA/Marshall Space Flight Center*

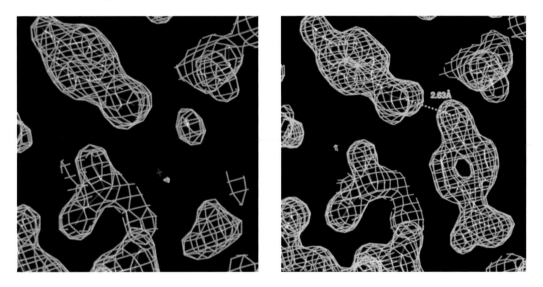

natively, if the nanostructures had some internal processing power, several ultrasound signaling devices could be applied to the body, so the nanostructures could triangulate their location and navigate to the tumor, rather as navigation is done on the earth using signals from several GPS satellites.

These types of artificial red blood cells or drug-delivery nanostructures have not been built yet "because molecular nanotechnology today is still theoretical," Merkle cautioned. "There's lively debate about how long it might be before such capability is available. It's probably measured in decades. It's not going to be in a year or two—but it's not going to take centuries, either."

Predictions: Rip Van Winkle, Call Your Doctor

Various science fiction stories over the last half of the twentieth century have been premised on the idea of suspended animation: a hypothetical process of freezing human beings, perhaps for decades, for travel over interstellar distances. The theory is that, at their destination, they are brought back to normal body temperature to resume their lives with the vigor of the age at which they were suspended.

Molecular nanotechnology may make cryonics—suspended animation—reality later in the twenty-first century, predicted Merkle.

"The freezing process itself is not benign, and causes tissue damage," he explained. "Today's medical technology is not able to reverse such tissue damage. All it can do is provide support to help tissues repair themselves. But if tissues stop metabolizing, physicians are helpless; they can only stand by and watch the tissues decline and lose function.

"But if, decades from now, we have medical technology based on molecular nanotechnology—the ability to rearrange the molecular structure of damaged tissue—it should be possible for an external repair technology to fix tissue that has stopped functioning. Thus, it should be possible to reverse tissue damage that is now regarded as unthinkable to fix, even frostbite. Ultimately, we'll be able to replace essentially all tissue in the human body except the brain," he suggested. (The brain poses a unique challenge, because that's where memories reside, and replacing brain tissues would cause amnesia.)

Teresa Brittain, Delta Flight Attendant, simulates an in-flight sudden cardiac arrest emergency and uses an automatic external defibrillator to respond. Precision electronics, and a better understanding of the heart's electrical pathways, is allowing the wider use of defibrillators by persons with less training than paramedics.
PR NewsFoto and Delta

RAY KURZWEIL: When computer intelligence exceeds human intelligence, what will it mean to be human?

RAY KURZWEIL/KURZWEIL TECHNOLOGIES

"Once computer intelligence matches human intelligence, it will quickly soar past it," said Ray Kurzweil, an artificial intelligence pioneer who is chairman and CEO of Kurzweil Technologies Inc., in Wellesley, Massachusetts. One reason he cites is "the inherent speed of electronic circuits, which are 10 million times faster than human neurons." Another reason is the ability of computer-based intelligence to transfer its knowledge and learning rapidly to another machine, something that humans can do only very slowly.

From observing current trends and rates of change in hardware and software, Kurzweil projects that "by 2029, computer intelligence will be more powerful than the human brain"—well within his own anticipated lifetime.

Now, if such tissue-repair capability should become reality, "that gives us reason to think that future technology could reverse the damage caused by freezing a human being," Merkle said. "You would restore the tissue structure at low temperature" before gently warming and reviving the person. The person could thus awaken a century from now in a world with vastly advanced medical capabilities—a boon for people who are considered terminally ill by today's medical standards, but whose illnesses might be curable in another 100 years.

"I'm a member of Alcor [based in Scottsdale, Arizona; *http://www.alcor.org*], which is an organization that freezes people," said Merkle. "Already there are about

At that point, he feels, one of the major difficulties facing people will be "getting used to the presence of nonbiological entities that are more intelligent than humans."

By intelligence, Kurzweil includes "all the diverse and subtle ways in which humans are intelligent—including musical aptitude, creativity, physically moving through the world, and even recognizing and responding to emotion," he said. "Will they be conscious?"

'By 2029, computers ... will seem to be conscious. They will *claim* to be conscious. And they will be very convincing.'

he asked. "They will seem to be conscious. They will *claim* to be conscious. And they will be very convincing" to any objective test.

But, he pointed out, there exists no objective test that can absolutely determine consciousness. "Science is about objective measurements and logical implications," he explained. "But the very nature of objectivity is that you can't measure subjective experience—you can only measure correlates of it, such as behavior. You cannot penetrate the subjective experience of another entity."

As a result, he continued, "we simply *assume* other humans are conscious. But there is not a consensus in humans about the consciousness of other higher non-human living beings that are known to be intelligent,"—such as primates, elephants, dolphins, and whales—and their ability to use tools and to possess sophisticated means of communication. "I feel my cat is conscious. Others say 'no, it's just a dumb beast operating by reflex.' Inability to know for sure is at the foundation of much misunderstanding and

The Lasette laser finger perforator is a revolutionary device designed for drawing blood from fingertips in a nearly painless manner for glucose testing. It utilizes a laser beam that penetrates the skin to obtain a capillary blood sample, instead of traditional steel lancets that tear the skin. It is the first medical laser device ever to receive FDA clearance for consumer use.
PR NewsFoto and Cell Robotics

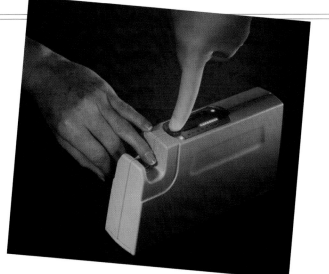

> **'There is no longer going to be a sharp distinction between humans and machines.'**

conflict," he said, pointing to entire races of people and other species that have been oppressed or exterminated by conquering invaders and settlers making similar arguments.

The question of consciousness and humanity is further complicated, he said, because in 20 or 30 years "there is no longer going to be a sharp distinction between humans and machines," he said. "We will put intelligent entities inside our bodies and brains to communicate with our environment." Blind persons, for example, may have a completely artificial visual cortex in their brains, just as some deaf people today have cochlear implants for doing auditory processing, and some Parkinson's disease patients have had neural implants installed in their brains to restore function. Before the end of the twenty-first century, he projected, "we'll be able to replace portions or even all of the brain." Even people without disabilities may elect to have neural transplants to augment their capabilities, he added—to the point where, in some individuals, "the balance between human and machine is dominated by nonbiological intelligence."

Moreover, by the end of the century, there may even be individuals "with biological origins who are in a nonbiological substrate," Kurzweil speculated. There currently exist invasive ways of scanning and mapping neurons in a brain and their interconnections, although not yet with sufficient speed to scan an entire brain. Noninvasive techniques for doing the same thing are more difficult, but progress is accelerating, he said. "Some people might choose to scan their loved ones into a computer so they can visit with them after their carbon-based body has died," Kurzweil suggested. "Or some peo-

30 people suspended in stainless steel Dewars, and another 450 signed up—including such notables as Marvin Minsky [artificial intelligence pioneer at the Massachusetts Institute of Technology] and Bart Kosko [optical computing pioneer at the California Institute of Technology]. You sign up in advance and pay for it with a life insurance policy. And when you're at risk of dying, you contact Alcor," Merkle explained. "They have to wait until you're legally dead before they can begin the freezing process—otherwise it's considered killing a person."

But Merkle expects that today's definition of "legal death" won't necessarily be regarded in the future as actual medical death. "Future medical technology based

ple might be motivated for their own longevity to scan their own brains" into a nonbiological entity. "Such an entity would have all your memories, your mannerisms, your creativity, your personality. But would it be *you*?" he asked. "Would it be conscious?"

Would it be entitled to all civil rights?

Would dispensing with it be considered murder?

Would it be human?

Although Kurzweil did not answer directly, he referred to the imaginary dialogues that conclude each chapter in his latest book, *The Age of Spiritual Machines: When Computers Exceed Human Intelligence* (New York: Viking, 1999). "Molly, who works for the Census Bureau, runs into this very issue," he said. "And eventually the bureau concludes that its requirement that one must demonstrate one's biological basis to be counted is unimportant."

◆

Ray Kurzweil is the inventor of a number of commercial firsts: a print-to-speech reading machine for the blind, an omni-font optical character-recognition (OCR) system, a text-to-speech synthesizer, a computer music keyboard capable of accurately reproducing the sounds of a grand piano and other orchestral instruments, and a large-vocabulary speech-recognition system. He founded, built, and sold four companies, all bearing his name. He is the recipient of the 1994 Dickson Prize (Carnegie Mellon University's top science prize), the Grace Murray Hopper Award from the Association for Computing Machinery, and nine honorary doctorate degrees in science, engineering, music, and humane letters. He is also the author of several books, including The Age of Intelligent Machines *(Cambridge, Mass.: Massachusetts Institute of Technology Press, 1990).*

on molecular nanotechnology should be able to reverse freezing injury, and therefore be able to revive the person. Death, by definition, is irreversible. If you can revive someone, they weren't actually dead."

To Merkle, the risks are well worth taking: "We know what happens to the control group. I'd rather be in the experimental group." And he has every confidence he'll wake up decades or centuries hence—and probably, he thinks, to a society that would be pretty nice: "First, they'd have made major advances in technology, otherwise they couldn't wake me up. And second, they'd value human beings, otherwise they wouldn't have bothered to wake me up." ◇

CHAPTER 8

TRANSPORTATION

AUTOMOBILES ARE ONE OF THE MOST pressing demands on environmental resources today, and will become even more so as developing nations acquire prosperity and Western-style tastes. Victor Wouk, president of Victor Wouk Associates in New York City, sees hybrid electric vehicles—those with internal combustion engines as well as batteries—paving the way for the ultimate adoption of pure electric vehicles. William F. Powers, vice president of research at Ford Motor Co. in Dearborn, Michigan, who started his career working on the guidance system for the Saturn V moon rocket, believes digital and power electronics will continue to "guide" the development of the automobile.

In time, automobiles may follow the lead of trains and aircraft in incorporating accident data recorders to determine the cause of a collision, says Linda Sue Boehmer, owner of the independent consulting firm LSB Technology in Clairton, Pennsylvania, who has been helping rail systems employ more embedded electronics. Meanwhile, the air traffic control system is about to move from an era of train-like dispatch and control to the freedom of the automobile, reports George L. Donohue, who is at George Mason University in Fairfax, Virginia, working to develop the first program in air transportation system engineering.

Left: This is where everyone wants to be, and what few people want to see change, in transportation. The driver's seat has remained remarkably unchanged for more than a half-century and is likely to stay that way as increasing numbers of people want the independent mobility that automobiles offer best. This is the front seat of a Toyota Prius, part of the new generation of gas-electric hybrids that will help lead the way to all-electric vehicles. *Toyota*

Electric and Hybrid Automobiles

"All energy used on Earth is derived from sunlight, except a small percentage from geothermal, tidal, and nuclear power," said Victor Wouk Associates president Victor Wouk. "The fossil fuels to power the 900 million vehicles that we can expect by 2020 took millions of years to form from plants in the pressure cooker of gravity and other natural forces in the earth." The demand for fossil fuels—50 percent for use in private automobiles—is far greater than the rate at which the fuels are being produced inside the earth. "Accordingly, we must think of a reasonably accessible alternative source of energy for individual road vehicle use."

Private automobiles must move away from gasoline-powered internal combustion engines for environmental reasons, too, he said—primarily to reduce the emissions of nitrous oxide (a principal ingredient of smog) and carbon dioxide (a greenhouse gas whose increase is implicated in global warming). Electric automobiles effectively still burn petrol, because for the most part, they are now recharged by fossil-fuel power plants. [For Wouk's thoughts about the tradeoffs people face, see his box "How much will we pay for freedom of movement?" p. 174].

The wireframe model in the foreground shows a power train quite different from what mechanics have come to expect over the last few decades. Ovonic nickel/metal hydride electric vehicle battery packs are installed on a dozen or more prototype EVs—both "purpose built" and conversion vehicles like the Solectria Sunrise and Chrysler TEvan. In 1996, a "purpose-built" Solectria Sunrise powered by Ovonic NiMH batteries traveled 375 miles on a single charge at the American Tour de Sol electric vehicle race. *Energy Conversion Devices, Inc.*

"The U.S. government recognized the evil health effects of auto exhaust before the gas-supplying countries did," Wouk said. "Beginning in 1968, Federal and California State governments began to legislate mandatory maximum limits to noxious emissions allowed in road vehicles." Although restrictions have become progressively tighter as technology has advanced, the worldwide increase in population and the demand for cars in developing countries threatens to erase the gains made. Since sport utility vehicles, classified as trucks and consequently not subject to the same restrictions as passenger cars, constitute 50 percent of the new vehicles on the road, air quality is deteriorating. Air pollution from cars is once again bad enough that "California has mandated that by 2003, 10 percent of new cars sold must be zero-emissions vehicles," Wouk said. And zero-emissions vehicle (ZEV) is usually interpreted as EV—electric vehicles.

"The EV has been the 'clean hope' of environmentalist since 1960 when Arie Hagen-Smit [then professor of biology at the California Institute of Technology in Pasadena] showed that the smog of Los Angeles was due to car exhaust reacting with sunlight," Wouk recounted. "The EV also became the 'oil-independent hope' of the conservationist during the 1970s oil crisis in the Middle East. Yet, despite an investment of billions of dollars, marks, francs, lira, pounds, and yen, the practical battery or fuel cell is still not here."

Wouk defines "practical" as the ability of the EV's battery to provide enough energy to drive 150 kilometers (93 miles) between 15-minute recharges, in a package whose form and price tag are comparable to fossil fuel cars. The initial lead-acid batteries were unacceptable in that they allowed only a maximum range of 100 kilometers (62 miles).

Recent advances in the more energy-dense nickel/metal hydride (NiMH) chemistry doubled that range to 200 kilometers—albeit without the air conditioner or other power-hungry accessories running—for a battery of the same weight. EVs might, at last, become truly interesting to consumers when they graduate to a lithium-based battery, which could yield a range of 350 kilometers. NiMH batteries—previously 10 times as expensive as lead batteries—now cost only four times as much. Lithium batteries are inherently expensive and are far more sensitive to overcharging.

Despite Wouk's optimism, Ford Motor Co.'s William Powers expects the public to be disappointed when advances in EV batteries do not duplicate the rapid advances of computers. "We assume everything electrical is on the same learning curve," he commented, "but batteries are not. They get a minimal percent improvement each year and are still very expensive."

Even in a laptop computer, where the battery is primarily moving bits of information rather than doing mechanical work (such as running disk drives), "you

VICTOR WOUK: How much will we pay for freedom of movement?

PHOTO COURTESY OF VICTOR WOUK

"Never mind all the jokes or profound analyses that cars are status symbols, sex symbols or whatever," said Victor Wouk, president of the consulting firm Victor Wouk Associates in New York City. "The fundamental attraction of automobiles is that they represent freedom—freedom to go where you want and when you want at a reasonable speed and at reasonable cost."

So what happens when 6 billion people pull up to the gas (petrol) station and the pumps are starting to run dry? "When the cost is too high, gasoline will become scarce and the freedom will be threatened. Hence, another source of propulsion energy will be required or mandated," he said.

That alternative energy source, Wouk believes, will be a battery or a fuel cell in an electric vehicle, with the hybrid vehicle—incorporating both a battery and an internal combustion engine or fuel cell—serving as a bridge that extends the life of fossil fuel supplies while battery or fuel cell technology slowly advances. Large volume manufacturing of batteries such as NiMH will be an important step in reducing the costs of batteries for EVs.

Ironically, the United States may be the last of the world's nations to make a transition

Ovonic NiMH battery packs can power an electric scooter like the one shown to travel more than 80 kilometers on a single charge, and be recharged to more than 60 percent of its capacity in about 15 minutes. These types of vehicles are an important mode of transportation in Taiwan, many European countries, developing nations in Southeast Asia and in India and China. Electric-powered vehicles can greatly reduce air pollution in the countries where conventional scooters are the primary source of nitrous oxide and hydrocarbon emissions.
Energy Conversion Devices, Inc.

to electric vehicles. Developing nations, with little existing infrastructure to change or write off, may become the most advanced from a transit standpoint.

"In China, few ordinary citizens have cars," Wouk pointed out. "They have bicycles. But if an electric-motor-assisted bicycle can make the difference between their living 6 kilometers from their job or 15 kilometers from their job—and living 15 kilometers from their job is a lot more attractive—then they will use an electric bicycle." And it's only a small step from wanting an electric bike to wanting a car of any sort.

While electric bicycles and scooters may seem heaven-sent in nations where people move on foot or crowded buses, Americans, who are accustomed to muscle cars, may prefer an airplane in the garage instead of switching down to a vehicle they perceive as tethered to an outlet.

Yet, Wouk feels the United States' adoption of electric vehicles is only a matter of time. "There will come a time when fossil fuels will be restricted to those applications where there is no viable alternative," Wouk said. At that time, customer expectations may drive EV technologies harder in the United States and other developed nations.

◆

Victor Wouk became interested in EVs in 1963 when he first investigated battery characteristics that might be better than conventional lead-acid. He recalls that when his first hybrid car was being tested at the Environmental Protection Agency at Ann Arbor, Michigan, the engineer in charge checked to make sure the engine was running because emissions were below background levels.

> '**When the cost is too high, gasoline will become scarce and the freedom will be threatened.**'

still can't go from New York to Los Angeles on a plane and have one battery work the whole way," Powers pointed out. Not only does battery technology still leave much to be desired, but the power demand to move the mass of a car is a far greater demand. Moreover, battery chemistry does not scale well, "that is, the bigger you make batteries, the tougher it is to make them work," Powers continued. "To move a 2,000-pound vehicle, you need a battery weighing nearly 1,000 pounds."

For the near term, he sees EVs as filling niche markets for short-range fleets operating in city centers where fossil-fuel vehicles are banned, or in limited situations, such as electric utility maintenance fleets.

The Prius hybrid passenger vehicle utilizes both a gasoline engine and an electric motor as motive power sources. It emits less pollution than ordinary cars, and it gets better gas mileage, which means less output of carbon dioxide. It starts on battery power, switches at speed to a combination of gasoline and battery, and brakes by using a generator to partially recharge the batteries. *Toyota*

Powers and Wouk agree, however, that one promising commercial all-purpose personal vehicle that will bridge the transition from gasoline-powered vehicles to EVs is the hybrid electric vehicle (HEV). The HEV—already marketed by a handful of companies, beginning with Toyota Corp. of Tokyo, Honda Corp. of Tokyo, and General Motors Corp. of Detroit, Michigan—is powered by a battery in stop-and-go city traffic. But once on the highway, a low-power gasoline internal combustion engine kicks in to maintain speed by making up for losses to friction.

Hybridization allows each power source to perform in the regime where it is most efficient. "The neat thing about electric motors is that they have such good low-end torque," Powers explained. "Rarely when you drive in cities do you need to get to 60 miles per hour. What you really want is performance from 0 to 15 miles per hour. And with an electric motor you can get your torque up so much more quickly, whereas there is a slight delay with most internal combustion engines."

Improving Accident Investigation

Regardless of the power plant, cars will crash. And automobile manufacturers are continuing to develop safety features that will make crashes more survivable for the driver and passengers. Safety belts remain the single best lifesaver, said Powers, followed by energy absorbing structures and airbags.

Auto makers are trying to make the car's handling safer, too. "Interactive vehicle dynamics use a rate gyro to detect when a car is rotating around its yaw [spinning] axis and then activate the antilock braking system and the engine management system to make small corrections to help a driver from getting into trouble," Powers explained. And for when the vehicle does hit, Ford and others are developing a sensor array that will determine what is in the passenger seat— a child or a bag of groceries—and limit or stop the deployment of an airbag so it causes no harm.

Moreover, after the debris stops flying, drivers may find themselves handing

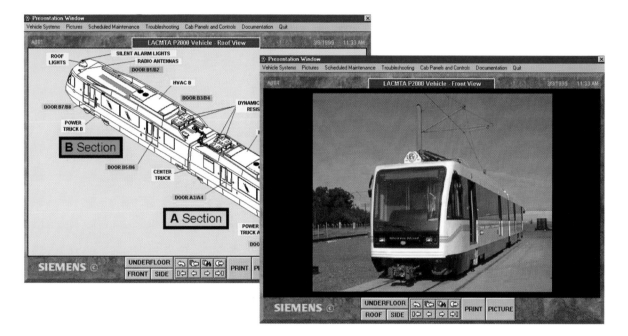

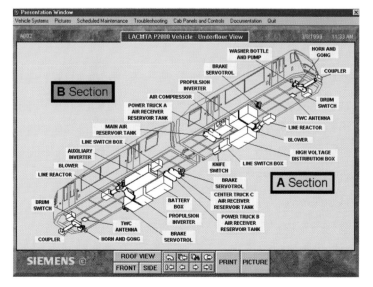

Screen captures from the interactive computer-based training program developed for the Los Angeles Metropolitan Transit Authority. These represent different screens and images during user interactivity, or demonstrate some of the many animated sequences used in maintaining the Green Line P2000 Light Rail Vehicle. *Siemens Transportation Systems, Inc. and the Los Angeles Metropolitan Transit Authority*

LINDA SUE BOEHMER: What is the potential of computer intelligence in mass transit?

"Just 15 years ago, I was spending a lot of my time educating the people in the rail industry about the potential of microprocessors," said Linda Sue Boehmer, a transportation consultant specializing in rail systems. "Now, microprocessor-controlled systems are everywhere. The next step is getting the maximum benefit from that built-in intelligence."

For years, "I had a vision of using a built-in intelligence for diagnosis," Boehmer said. "One of the biggest problems for the people who have to take care of mass transportation systems is fixing them. You hate to have a whole trainload of people stalled in a dark tunnel in the New York City subway." Because microprocessors now control everything from the optimal operation of motors to optimal comfort for passengers, however, they also "know" which parts of the system are doing their job—and which ones are about to quit. Thus, "once you have all that processing power, if something goes wrong

over the car's "black box" so a police officer can determine who hit whom. "New York City taxis have those cute little boxes that print out your receipt automatically," complete with mileage and trip time, pointed out LSB Technology's president Linda Sue Boehmer. "There is no reason why that box couldn't have a piece of flash memory in it and be hardened to survive a crash, and be connected to a couple of key intelligent systems on the car. I think in the next two decades we will certainly implement a crash recorder on any public transportation vehicle."

Already, the Massachusetts Institute of Technology, in partnership with Japanese-based Mitsubishi Electric's Information Technology Center America, both

or is about to go wrong, you can flag it" so a human operator can do something about it, said Boehmer.

Moreover, since the many intelligent systems can talk to one another via various networks, "you can collect information about any of the problems the systems are having," she continued. "You have the built-in diagnostic capabilities of each intelligent system being networked and available in a centralized place." The result? "On newer train cars, there is a central receptacle where you can plug in [a laptop computer], just like in the movie *Star Wars*, where R2D2 plugged in and was able to find the heroes in one of the trash compactors," Boehmer declared.

"We are just barely scratching the surface of that functionality," which to her, as a designer and specifications writer, is very exciting. In her view, the next challenge is linking onboard control intelligence, diagnosis capability, and test equipment with documentation to present technical information "in a way that can be [readily] assimilated by the people who have the care and feeding of that system."

She envisions the full integration of technical manuals and training techniques into the transportation system's information backbone, to provide the necessary information, instantaneously, in the most effective form to the people who need it. "Traditionally, the maintenance personnel have been provided with written manuals and training classes. But if you want to make an engineer cranky, just say, 'Now that you have designed this great widget, you have to write the manual for it and provide the training to the guy who is going to take care

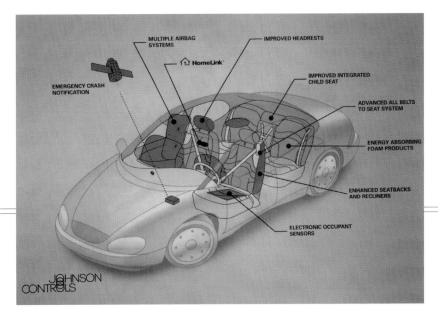

A wide range of electronic sensors and systems are being added to cars to enhance passenger safety. The new features do everything from correcting the driver's braking and steering to slowing or stopping airbag deployment to reduce injuries to small passengers. Also included in this design by Johnson Controls are an emergency crash notification system; advanced and repositioned headrests; and seats, headliners, armrests and other interior components that would absorb more energy in vehicle crashes. Above all, the systems are designed to be "transparent"; the driver is largely unaware of their presence. *PR NewsFoto and Johnson Controls*

'The next frontier is systems that are self-repairing.'

of it for 20 years.'" The traditional result, she said, has been manuals that are cumbersome to cross-reference and use, and training classes that are variable in quality and completeness, both of which could make mastery of the information tedious and slow.

Increasingly, however, documentation is automated, she said. "Now maintenance manuals are commonly provided on CD-ROM. The obvious next step is to take the World Wide Web and just hyperlink it" to the system's database so the user doesn't have to carry even the CD-ROM.

The next frontier, she speculated, is "systems that will be, to an extent, self-repairing." When a problem is discovered by the train's diagnostic systems, it could be radioed to a human operator or artificial intelligence at a wayside location; the wayside, in turn, could suggest several options for fixing the problem that the onboard "technology can actually implement autonomously. That's the exciting thing I can envision."

◆

Linda Sue Boehmer is owner of LSB Technology, Clairton, Pennsylvania. She earned a B.S. degree in electrical engineering from Carnegie Mellon University in 1974. She is past president of the IEEE Vehicular Technology Society and chairs its working group to establish standards on Monitoring, Diagnostics, and Event Recorders. She is a member and past officer of several technical and professional societies, including the Society of Women Engineers. A trailblazer in applying advanced technologies to the rail industry, she has been the author or subject of dozens of technical papers, articles and books and is a popular speaker on "high tech." She also is a private pilot, holding several ratings, and is a member of various aviation organizations.

in Cambridge, Massachusetts, have developed the equivalent of an aircraft's black box recorder for use in fleet trucks. The Personal Eye Witness, costing only a few hundred U.S. dollars, comprises a TV camera taking images five times a second and a microcomputer that is constantly recording over and over again in its memory. At the instant of a collision, the computer freezes the last 20 seconds and records for another 10 seconds, providing enough video to show what caused the accident and its immediate aftermath.

Meanwhile, the potential for diagnostic event recording has been on Fords and other brands of vehicles for almost a decade to help service departments deal

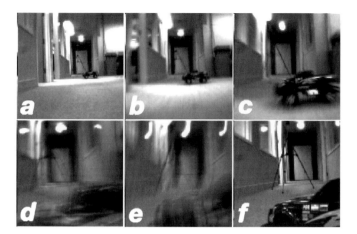

No more questions about who hit whom. The Personal Eye Witness, being developed at MIT, would record the last 20 seconds leading up to an accident and the 10 seconds immediately after an accelerometer was tripped. The camera only offers a 128 x 128-pixel view. But as shown by lab tests on toy trucks, that's enough to establish that the truck in the distance was parked, and the "driver" rammed it without stopping. *MIT, Mitsubishi Engineering Research Laboratory*

with intermittent problems that seem to disappear mysteriously in the presence of a mechanic. [For details about such systems, see Boehmer's box "What is the potential of computer intelligence in mass transit?" p. 178.]

Privacy issues, however, may slow the adoption of such systems into commercial use. Long-haul truckers resisted—and sometimes sabotaged—the installation of navigational satellite receivers that tracked their trips. They may also resist giving management a view from the front seat. "The biggest barrier against getting more intelligent trip recorders has been that the operators—the pilots, the train engineers, and the drivers—are nervous about Big Brother," Boehmer said, referring to the all-seeing computer in George Orwell's classic dystopian novel *1984*. In public transportation, "it has been an ongoing battle in the aviation industry because the U.S. National Transportation Safety Board is trying to increase the number of signals that are monitored and the manufacturers and owners are always trying to increase signals for predictive maintenance reasons." And passengers increasingly will feel that the carriers owe them the added degree of safety.

"I do think it makes you aware you are being watched," Boehmer added. "I think there is a question about privacy issues and often there is a tradeoff between privacy and public welfare: my right to drive or fly any old way I want and someone else's right to stay alive instead of being smashed into by me."

That being said, "something has to happen to bring insurance premiums and litigation back in line," Boehmer said. "This bit of technology could totally revolutionize the auto insurance industry" as well as encourage drivers to be safer, if they knew blame would be assigned with certainty. "If insurance companies say, 'If cars had this system we will give an X percent reduction,' we could probably get it into production pretty quickly," Powers agreed.

When you get ready to shop for a twenty-first-century car, though, don't look for the 1958 Disneymobile, the car that drives itself while the family relaxes

Satellite-based navigation systems aboard automobiles are highly popular in older nations where streets are poorly marked and among U.S. travelers going into cities they are unfamiliar with. Philips CARIN 520 Vehicle Navigation gives both visual and voice turn instructions to navigate the driver through unfamiliar streets and traffic, and helps the driver navigate to the desired destination. Such databases can be updated to avoid construction and high-crime areas. *PR NewsFoto and Philips Carin*

and chats. Cruise control may be as sophisticated as it gets. [For Powers' thoughts on limits to consumer interest in technology, see his box "Will cars ever have jet fighter controls?" p. 184]. "We have never had a problem making one car follow another car," Powers explained. "The controls for that are straightforward." But in fitting self-driving cars into the transportation infrastructure—including meeting the demands of juggling every vehicle getting on and off the road, and last-minute decisions to get a burger or visit the restroom—"you get to degrees of complexity and expense that just don't make a lot of sense."

An Airplane in Every Garage

While government agencies may dream of bringing air traffic control-like technologies to the highways, the aviation industry may be on the verge of gaining automobile-like autonomy. George Mason University's George Donohue views aviation history as falling into three epochs, each lasting about a third to a half of a century.

The first epoch, he said, started with the Wright brothers, who integrated the first successful system of strong, lightweight structures, aerodynamics, engine power, and control. It took off with the large-scale production of fabric and wood airplanes in World War I.

The second epoch began in the 1930s with the Douglas Aircraft DC-3, the first successful long-haul passenger liner, and continued to the end of the twentieth century. In it, aviation matured with all-metal aircraft, onboard electronics (including VHF radio and radar), and the jet engine—which yielded all-weather operation, jet transports that can fly above bad weather, and the modern air traffic control system.

"Now we are choking, coming to gridlock on the hub-and-spoke system that the airlines use," Donohue said. "I really see no way to avoid it. By 2010, we are probably going to have commercial air transportation that looks like our urban

Although short-haul airliners may lead growth, "bigger is better" will continue to drive long-haul airlines. Airbus Industrie Inc. of France announced plans in early 1999 to build a 600-passenger, double-decker jetliner. (The first-class deck of the Boeing 747 only extends one-quarter the length of the airframe.) *Airbus Industrie Inc.*

WILLIAM F. POWERS: Will cars ever have jet fighter controls?

"One of our rules of thumb is that when we put anything on a car that is technology driven, we make it as transparent as possible to the customer," said William F. Powers, vice president of research at Ford Motor Co. in Dearborn, Michigan. "We know that customers prefer systems that are easy and intuitive to use. They just paid 500 bucks for an option and now they have to take extra time to learn how use it? They are going to be doubly mad."

Powers spent the early part of his career working for NASA on the guidance system for the Saturn V rocket used to boost the Apollo astronauts toward the moon. Yet, now that "space age" technologies are becoming available on most models of cars, he foresees little interest for them where the driver would use them. A simple example is keyless entry, in which the car's lock is a five-digit numeric pad. "It would take you about a minute to learn how to program your own number," Powers said. "But most customers don't want to take

freeway system today: lots of delays, very uncomfortable, and a lot of slowdowns in the ability to move through the air."

In his opinion, the answer is not to make planes faster. "The airplane is not the choke point anymore," he pointed out. "Planes fly 450 knots [825 kilometers per hour] routinely, giving most flights less than 100–150 knots [185–275 kilometers per hour] average transit speed because of all the other delays in ground transportation times, air traffic control, and hubs." Getting to a destination faster also puts travelers' own body clocks out of whack so they have to sleep rather than being able to jump right into business or touring.

even a minute to learn about something that simple." So the dealer usually does it.

"Customer-discernible technologies are difficult to develop for a car," Powers said. "When most of us started driving in the 1950s or '60s, cars had a steering wheel, four wheels on the road, and analog displays. Just think of all the revolutions that have happened in the world since then: computers, atomic energy, airplanes went from propellers to turbine engines. Yet cars still look the same. They still have a steering wheel, four wheels on the road, and analog displays. In the '80s, when digital electronics were really coming on, we thought that displays would go totally digital. But customers like dials.

'Most customers don't want to take even a minute to learn about something [on their car].'

"There are over 700 million vehicles in use across the world today. And every auto company in the world is out checking these people, doing market research, trying to come up with something new that will grab customers."

A potential grabber, once found only in military aircraft, is night vision. Cadillac will soon add it to its top-line cars, but Ford will wait to see how the consumers accept it. Customer resistance has, for example, been the reason that fighter-pilot-style head-up displays—where images of dials are superimposed on the windshield—have not caught on a decade after being introduced to consumers.

In fact, one of the most frequent complaints that automobile manufacturers hear is about the wide variety of display and control arrangements—an increasing problem as more people travel on business and have to adjust to rental cars that are different from what they have at home. Such complaints are one reason that auto makers spend so much

"I think the movement is going to be not to a bigger mass transportation system," Donohue said, "but to smaller airplanes in a personal transportation system." This, in his view, will be the dawn of aviation's third epoch.

The first step will be a free flight system in which airline pilots are free to fly as straight a line as possible from point to point, with appropriate detours for weather and, as directed by ground control, around each other. "The microcomputer, extremely accurate clocks, satellite navigation, and satellite and digital communications are combining to give us a satellite-based air traffic surveillance system," Donohue said. "If you know where you are very accurately, and you have a digital

'We are trying to blend technology with what customers really want.'

time "hunting for what the customer really wants," said Powers. "We are trying to blend technology with what customers really want, and then make the technology as transparent and as low cost as possible."

Powers does see modest success in the United States for onboard computers using satellite navigation and digitized maps. "They are very popular in rental cars," he noted, although people don't want to pay the expense for them for their regular car—nor are they necessary in the United States, because the highways are so well-marked. Europe and Japan, however, have countless, unmarked paths and alleys that have been pressed into service as streets, so getting lost is much easier.

"And those countries have a lot of traffic congestion," Powers said. "In Japan, when you're stuck in a traffic jam, drivers like to stick the car in park and watch TV on the navigation system."

◆

William F. Powers is vice president of research at Ford Motor Co. He earned his B.S. at the University of Florida in aerospace engineering in 1963 and his master's and Ph.D. at the University of Houston in aerospace engineering in 1966 and engineering mechanics in 1968, respectively. He was a professor of engineering at the University of Michigan for 13 years before joining Ford in 1979. He has authored more than 50 refereed technical articles, and serves on the board of directors of the Intelligent Transportation Society of America.

communications capability, you can broadcast your location, speed, altitude, and intent much more rapidly than radar. It is a major step beyond radar that provides situational awareness in the cockpit." It started phasing in with Pacific Ocean routes in 1994, and will become widely implemented by 2005. [For Donahue's views on human limitations to the adoption of new systems, see his box "Can we overcome our fear of flying?" p. 188.]

Free flight will bring about, at long last, the post-World War II dream of an airplane in every garage. After the war, it was just wistful thinking by manufacturers hoping to maintain their wartime production levels. The required flying

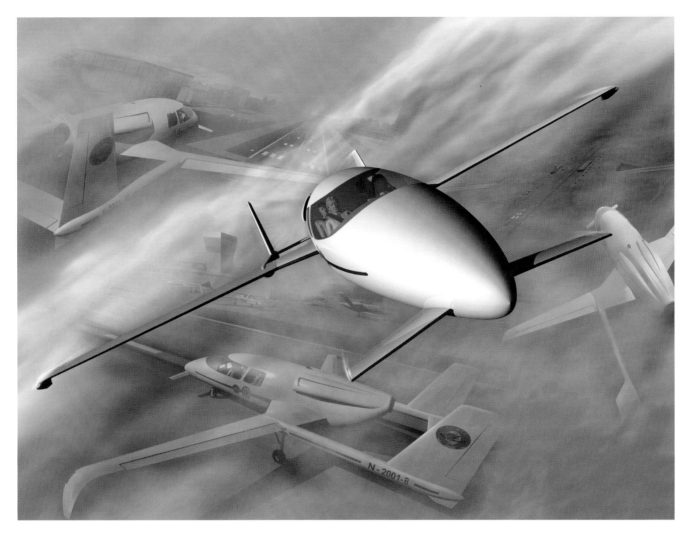

Computer-generated artist's concept of a consumer aircraft evolved from the Advanced General Aviation Technology Experiment (AGATE) project at NASA, which could lead to an "airplane in every garage" in the twenty-first century. Autopilot technologies, combined with advanced displays and computers, will reduce much of the workload and skills requirements for piloting aircraft even in the worst weather. This design employs advanced composite structures in wings and tail structure that look too frail to fly. *NASA/Langley Research Center*

skills were too high for the mass populace. But those requirements may be easing. Donohue predicts that, thanks to advanced electronics and free flight, a pilot's license will be only about twice as challenging to obtain as an automobile driver's license—and no longer terribly expensive, since a lot of learning can now be done with flight simulators running on desktop PCs. Moreover, ownership of a private plane will be within reach of mid-sized businesses or individuals in the upper class who can afford a Lexus (Toyota's top luxury car, which costs about U.S.$50,000, or the same as an inexpensive home).

The Lexus comparison is especially apt since airplanes today are still made

GEORGE L. DONOHUE: Can we overcome our fear of flying?

"People have to become more comfortable with the thought of flying" in order for aviation to expand now that advanced electronics can do most of the flying, said George L. Donohue, visiting professor of air transportation technology and policy at George Mason University in Fairfax, Virginia. "People are one of the barriers to this vision becoming a reality. This is cultural; it is not a technical barrier any more."

That sounds surprising, since the United States has been one of the world leaders in aviation and is one of the few nations where aviation has all but killed rail transportation.

"It goes to the psychology of fear and control," Donohue continued. "It is probably initially why people were afraid to get in the car. You had to be a brave person to get in

the way they were 50 years ago, Donohue said, while automotive technologies have sprinted ahead of aviation technologies: "Automobile manufacturing exceeds aerospace standards. Something high-end like a Lexus has more sophisticated electronics than a private airplane. It has very sophisticated composites and manufacturing techniques, all that far exceed anything in aviation manufacturing. And you get a reliability with the electronic parts in a car that exceed that which you have in avionics in an airplane." Thus, low-cost aircraft may require that the manufacturers look beyond techniques they have used for the past half-century.

With projects such as the Advanced General Aviation Technology Experiment

a car and go over 20 or 25 miles an hour, because only horses were safe. You knew how to ride a horse, but you didn't know how to drive a car. People know how to drive a car today but they don't know how to fly an airplane. A psychologist would say they fear planes because they don't think they are in control. They are afraid that if the airplane gets in trouble, they can't go up and fly the plane—whereas they think, 'If I'm sitting in the right seat and the driver has a heart attack, I know how to drive a car.'"

Headline news also is a factor. Every time a large commercial airliner goes down, it gets intense news coverage. Such coverage "plays to an irrational fear," Donohue explained. "Dying from an airplane collision is perceived as being worse than dying from an automobile collision. I never understood that. Dead is dead. Being ripped apart is being ripped apart. More people are afraid of dying in an airplane than they are of dying in a car, although more people die in cars. Yet, we have a much higher demand for safety in air travel than we do in ground travel. Moreover, when calculated according to the percentage of hours you are in contact with the risk, air and ground transportation have about the same mortality rates."

Zero flight fatalities has been set as a goal by the U.S. Federal Aviation Administration. While achieved in 1998, it is unreasonable to expect such a goal to be reached every year. Here, rescue team members survey the wreckage after Korean Airlines flight 801 crashed on Guam on August 6, 1997. *Petty Officer 3rd Class Michael A. Meyers, U.S. Navy*

'[People] are afraid that if the airplane gets in trouble, they can't go up and fly the plane.'

Donohue believes the fear of flying is driven partly by the U.S. Government. U.S. Secretary of Transportation Frederico Pena even decreed that he would tolerate no fatalities in commercial aviation (excluding charters, business, and private aircraft)—and, in fact, aviation had no deaths on commercial flights in all of 1998.

But on the highways, Pena would tolerate 40,000 deaths a year—the equivalent of 10 major air crashes a month. At the same time, the public demands eased highway speed limits, helmet laws, and airbag requirements. "These movements are 180 degrees out of phase with each other," Donohue commented.

◆

George L. Donohue spent four years as Associate Administrator of Research and Acquisition at the U.S. Federal Aviation Administration, where he helped restructure the National Airspace System. He is a Fellow of the American Institute of Aeronautics and Astronautics, and a member of the Experimental Aircraft Association. He also has published over 50 articles and reports and holds a patent on an oceanographic velocity sensor. Among the various awards and recognition he has received during his career was the Secretary of Defense Meritorious Civilian Service Medal in 1977, and he was named one of the top 100 decision makers in Washington, D.C., by the National Journal *in 1997.*

(AGATE) being pursued jointly by the aviation industry and the U.S. National Aeronautics and Space Administration (NASA), Donahue's third epoch in aviation is about to take off. "As people see this new airplane [using AGATE-derived technologies] and how easy and safe it is to fly, we are going to get the same type of excitement building up that we did in the '50s with an 'airplane in every garage.'" Initially the aircraft will be fixed wing, thus tying them to small airports. In time, the technologies will expand to make vertical takeoff aircraft more widely available, allowing them to leave and land on downtown heliports.

"It will move down to the Chevy class. That might take 50 years. When we

Einstein called it "spooky action at a distance." Scientists today, like Mr. Spock of *Star Trek*, call it "fascinating." Physicists in Switzerland (soon followed by others in Germany and the United States) in late 1998 demonstrated "quantum teleportation" in which two photons are separated. At a distance, a change in the properties of one is reflected in an instantaneous change in the other. In effect, it was communication at speeds faster than light. The apparatus is practical only for laboratory experiments—for now. *IBM*

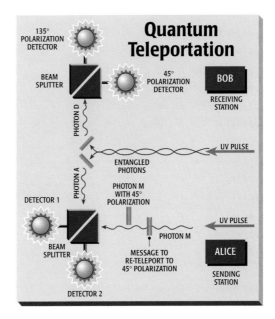

get to that point, it might be the end of the third epoch of flight," Donohue said. "I have no idea what is going to be the fourth epoch."

Predictions: 'Beam me up, Scotty'

Perhaps the fourth epoch will skip directly to the ultimate for transportation, teleportation—the sending of the physical object itself at the speed of light, with no need for raw materials or manufacturing apparatus.

In 1997 and 1998, scientists in Europe and the United States announced that they had achieved what the scientists claimed to be "the first bona fide teleportation." First, teams in Innsbruck, Austria, and Rome, Italy, said they teleported individual photons (particles of light). But in October, 1998, a team led by Jeff Kimble, a physics professor at the California Institute of Technology, reported in the journal *Science* that they had teleported an entire beam of light across a laboratory bench. Using a little-understood physical property called quantum entanglement— once described by Albert Einstein as "spooky action at a distance"—they created two entangled beams of light to carry information about the quantum state of the third beam. The first two beams were canceled in the process, but the third successfully transmitted its properties over a distance of about a meter.

Although all the investigators worked only with light, Kimble thinks teleportation could be applied to solid objects, such as individual atoms. The question not yet understood is whether the object thus teleported is the original object or simply a perfect replica. But Kimble left open the possibility that the transporters of the television and movie science-fiction series *Star Trek*—which beam people and objects between the ship in orbit and a planetary surface—could one day be reality. ◇

CHAPTER 9

EXPLORATION

THE EARLY TWENTY-FIRST CENTURY will see a radical expansion in human ability to reach remote places, from other planets to the ocean bottom to the innermost depths of the atom. Some of this exploration may be done by humans using "virtual" bodies. Moreover, just as settlers and ultimately tourists have rushed in where explorers once trod and died, by the end of the century ordinary people may be living in space and ocean communities.

These are a sampling of opinions from several individuals probing the frontiers of exploration today. Freeman J. Dyson is a physicist who has explored concepts for exotic rocket propulsion systems and methods for the search for extraterrestrial intelligence. Burt Rutan is an engineer pioneering the design of innovative aircraft, including *Voyager*, the first manned aircraft to circle the Earth without refueling. Joseph R. Vadus is an electronic and ocean engineer and former director of manned and unmanned undersea technology research at the U.S. National Oceanic and Atmospheric Administration (NOAA). And Robert Zubrin is an astronautical engineer best known for proposing a quick, low-cost strategy for becoming self-sufficient on the planet Mars.

Left: An artist's concept depicts a Lightcraft as it is hoisted on blasts of air superheated by a high-energy laser coming from the left. Lightcraft acts as a ramjet, with air entering through small slots in the skirt and then pushing against the skirt as the air expands. Initial use early in the twenty-first century would be to launch small satellites, but it has the potential for payloads weighing a few hundred kilograms.
NASA/Marshall Space Flight Center

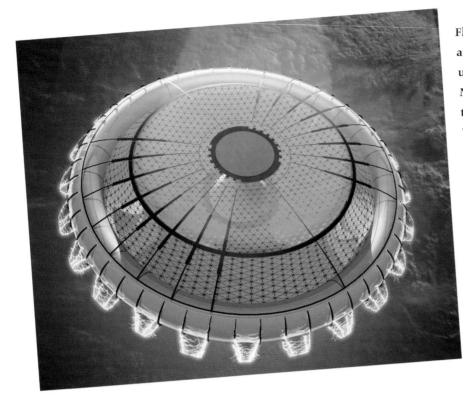

Flying saucers, anyone? Students at Rensselaer Polytechnic Institute, under the direction of Leik Myrabo, are going a step beyond the laser-powered Lightcraft. The disk is made of silicon-carbide that is transparent to high-frequency microwaves transmitted by an orbiting power station. Inside the disk, and wrapped around a crew compartment, is a parabolic mirror that focus the microwaves so they heat the air, thus propelling the Lightcraft.

John Frassnito & Associates for Rensselaer Polytechnic Institute

Lighting the Way off the Earth

"Laser propulsion, if it works, has the promise of being about 100 times cheaper than chemical rockets" for getting people from Earth into space, claimed Freeman J. Dyson, professor emeritus of physics at the Institute for Advanced Study at Princeton and author of the best-seller, *Disturbing the Universe*. "In principle, cheap transportation from the ground into space would transform the whole business. It will take 50 or 100 years, of course, to build all the systems and the infrastructure (such as space ports, hotels, and factories) to go with it. But if it works, space travel could become like present-day intercontinental air travel—something that is routine."

Laser propulsion was first proposed more than three decades ago by Arthur Kantrowitz of A.D. Little, Inc., of Cambridge, Massachusetts. Laser propulsion uses lasers to transmit energy that expands a gas to push the spacecraft up. It's a rocket without combustion. The test model, called Lightcraft, uses a ground-based high-energy laser developed in the Strategic Defense Initiative ("Star Wars") program and a polished metal craft resembling a bundt cake pan. Lightcraft's parabolic base cavity focuses the laser light on a region immediately under the craft and superheats the air, causing a small explosion that pushes the craft upward. As the laser pulses, the Lightcraft rises ever higher, riding upward on the beam of light. As the air gets thinner, the Lightcraft supplies its own gas to continue the process.

The working model, only 15 centimeters (6 inches) in diameter and weighing 50 grams, was developed by Leik Myrabo, associate professor of mechanical engi-

A Roton rotary rocket lifts off at dawn at Edwards Air Force Base, California. Rotary Rocket is one of several companies embracing private funding and (sometimes) radical approaches to open the space frontier. The Roton takes off like a conventional rocket, re-enters the atmosphere somewhat like a missile warhead, then deploys helicopter rotors to provide controlled flight to a soft landing. If all works as planned, the Roton will be "turned around" and ready for its next flight in two or three days. A two-person crew rides in a small, windowed ejection capsule near the base. *Rotary Rocket, Inc.*

neering at Rensselaer Polytechnic Institute, Troy, New York, with funding from the U.S. National Aeronautics and Space Administration (NASA) and the U.S. Air Force. In October 1997, it made a 4.3-meter–high arc in a 2-second flight. In Dyson's opinion, the feat was comparable to the Wright Brothers' demonstration of heavier-than-air flight in 1903. By early 1999, Lightcraft was flying as high as 36.5 meters (120 feet), a limit set by a crane-held light shield.

"Now it is a question of how you get from the equivalent of the Wright Brothers to the 777 jetliner," which in aviation history took almost a century. A full-sized Lightcraft, perhaps a few meters in diameter and capable of launching humans into space, would be propelled by a gigawatt laser. "The spacecraft itself is the problem," Dyson admitted. "We know how to build big lasers but we do not know how to build laser-driven engines. But at least we did make the first step off the ground, which in some ways is the hardest." In time, he foresees laser-driven light-

FREEMAN J. DYSON: How can we further explore the 'microverse'?

RANDALL HAGADORN

"I would expect as soon as you start looking at smaller scales, you'll find things you don't expect," said Freeman J. Dyson, professor emeritus of physics at the Institute for Advanced Study in Princeton, New Jersey. "I would be very disappointed if there weren't any big surprises in particle physics."

What would it take to look at ever-smaller scales? "New ideas," Dyson replied. "Existing particle accelerators have reached the point of diminishing returns." Penetrating further into the world of subatomic particles "will take more powerful tools than scientists now have available."

"Essentially nothing has changed since E. O. Lawrence built the first cyclotron in 1931," Dyson explained. "We still use the same principle for accelerating [particles]: radio

sails accelerating mankind's first miniaturized payloads into interstellar space. [For Dyson's thoughts on another application of high-power lasers—exploring the innermost secrets of matter—see his box "How can we further explore the 'microverse'?" above.]

Myrabo is leading engineering students at Rensselaer Polytechnic Institute in designing a version of Lightcraft that would be powered by microwaves beamed from a power station in orbit. The helium-filled, 20-meter-diameter, 12-person Lightcraft would use the microwave energy in several different forms to propel the craft. Ion-propulsion engines electrify and heat the air on one side of the buoyant craft

waves—alternating electric and magnetic fields—on the scale of meters or centimeters in wavelength accelerate the particles" around in a circle or along a straight track. That principle of acceleration "sets a fundamental limit to how many volts per meter you can apply, which is about 100 million."

> **'Existing [subatomic] particle accelerators have reached the point of diminishing returns.'**

The irony of particle physics has been that looking at ever smaller parts of the universe has meant building ever larger accelerators because of the energies required to pry open new secrets. "That means that when you want to go to higher and higher energies, the machines must get bigger and bigger," Dyson said. "And that is why we have essentially run into a practical economic wall," because now to penetrate to higher-energy particles, accelerators have to "be larger than 10 kilometers in size," he said. Economics, among other reasons, is what forced the cancellation of the 27-kilometer-diameter Superconducting Supercollider in Waxahachie, Texas, which would have been the world's largest particle accelerator.

But Dyson does not think it has to be that way. In fact, he thinks, it may be possible to build tabletop accelerators that perform breakthrough physics.

"To me, lasers look to be very promising as accelerators," Dyson said. "Laser light has much larger electric fields" than ordinary cyclotrons, he pointed out. "Ordinary lasers—things that costs a few hundred dollars—can produce fields of 100 billion volts per meter rather than just 100 million. So they are about a 1,000 times stronger as far as the electric fields are concerned. In principle, you could build accelerators a thousand times smaller, if you could just invent some clever way of doing it." In theory, a laser

to push it in the other direction. Pulsed detonation engines would heat air and expel it through slots around the perimeter of the disk. Finally, the microwaves would be focused into the air to form a bubble of hot gas that would serve as a nose cone during supersonic flight.

For the early twenty-first century, though, getting off the ground will continue to involve conventional rockets. But even private ventures are getting into the business. One is Scaled Composites Inc. in Mojave, California, the company founded and headed by Burt Rutan, a pioneer in aircraft design. Rutan is designing and building the structure of the innovative Roton being developed by Rota-

accelerator only 30 or 35 meters long could accelerate particles to energies comparable to those attainable with the Superconducting Supercollider. "We don't yet know how to do it, but I would be surprised if this isn't possible."

There are, of course, technical challenges. "The real problem is that even though the laser fields are very strong, the regions of space in which they exist are very small. So you have to have your particles in the right place at the right time on a microscopic scale. It is not clear how you could possibly arrange that," Dyson said.

"But, when people start to think about a problem, it generally happens that somebody figures out something clever," he added. "Fifty years ago, nobody would have believed that you could build computer chips the way we are doing it now. So I would not say it is not going to happen." In fact, he added, "I would say it looks promising."

'To me, lasers look to be very promising as accelerators.'

◆

Freeman Dyson is best known for his speculations on contacting extraterrestrial intelligence—and believes that we don't yet know enough to conclude one way or the other whether it exists. He worked on the Orion Project, a concept for using a rapid series of nuclear explosions to power an interplanetary spaceship. He is a fellow of the Royal Society and a member of the U.S. National Academy of Sciences. Dyson also is a celebrated writer who speaks with eloquence to scientists and laymen alike.

ry Rocket, Inc., in Redwood City, California. The Roton will be a single-stage-to-orbit launcher that relies on conventional liquid rockets to reach space. But unlike any earlier return to Earth, the Roton will use turbine-powered helicopter blades whirling at hypersonic speed to help it land.

The Roton "is an example of something high-risk, courageous," Rutan remarked. "If people had only tried one airplane concept every 10 years, we wouldn't have gotten very far. But hundreds of people tried different approaches. I would say that Roton is one of those opportunities. If a rotor for recovery for a spaceship is a good idea, you are not going to find that out just studying a computer. You've

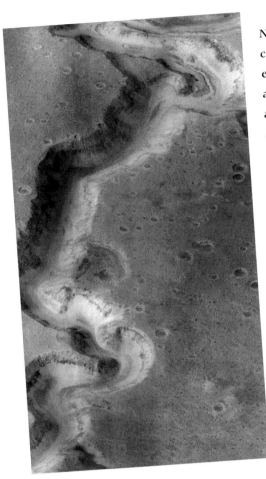

Nanedi Vallis, one of the Martian valley systems cutting through cratered plains in the Xanthe Terra region, provides additional evidence that Mars once had running water. The picture covers an area 9.8 by 18.5 kilometers (6.1 by 11.5 miles), and features as small as 12 meters (39 feet) can be seen. The canyon is about 2.5 kilometers (1.6 miles) wide. The origin of this canyon is enigmatic: some features, such as terraces within the canyon and the small 200-meter (660-foot) wide channel, both visible at the top of this photo, suggest continual fluid flow and downcutting. Other features, such as the lack of a contributing pattern of smaller channels on the surface surrounding the canyon, box-headed tributaries, and the size and tightness of the apparent meanders, suggest formation by collapse. It is likely that both continual flow and collapse have been responsible for the canyon as it now appears. *NASA, Jet Propulsion Laboratory, and Malin Space Science Systems Inc.*

really got to go out and try it." [For Rutan's ruminations on his motivations, see his box "Why are humans driven to explore?" p. 200.]

No matter the propulsion system used to break the bonds of the earth's gravitational field, unconventional concepts are also being advanced to continue the journey through the vacuum of space. Electricity is the key in many schemes. The 486-kilogram *Deep Space 1* spacecraft, launched on October 24, 1998, for a July 29, 1999, flyby of asteroid 1992 KD, is demonstrating propulsion using electrostatic repulsion of xenon gas. "Solar-electric ion propulsion is very good for long-range voyages," Dyson explained. "Once you get out of the earth's gravitational field, all you need is low thrust for long periods of time."

For faster trips with heavy payloads, including manned missions, the Jet Propulsion Laboratory in Pasadena, California, Princeton University, and the Moscow Aviation Institute are studying Lorentz Force Accelerators. These produce thrust when a charged particle—such as vaporized lithium metal—moves through an electric field and a magnetic field.

Onward to Mars

Robert Zubrin, president of Pioneer Astronautics, Inc. in Indian Hills, Colorado, is energetically promoting a cheap, quick, and simple way to get humans to the planet Mars within a decade. With David Baker, he devised the Mars Direct approach after NASA's response to U.S. President George Bush's proposal for a

BURT RUTAN: Why are humans driven to explore?

SCALED COMPOSITES

"I think we need to explore. I think if we stop exploring that we will run into mediocrity and we all will get bored," declared Elbert "Burt" Rutan, an aviation pioneer who has stretched the limits of what we think aircraft can do. "And we won't make progress.

"Exploration is one of the few things that you can look at and say, 'This is what makes us different than the animals,'" Rutan continued. In 1986, Rutan's twin-engine *Voyager* aircraft made the first and only non-stop, round-the-world flight. If he has his way, he may stretch the limits for spacecraft, too. "It is not just coincidence that the size of our bodies being halfway between the smallest thing we know to exist and the largest thing. I think that really tells you that there is a lot more to look at—whether that means examining things smaller than the smallest particle than we can now, or looking to the edge of our universe."

With technology probing into the very essence of matter, the edges of the universe, and covering the globe in detail, what is left to explore in the twenty-first century?

"Certainly there are still horizons for exploration," Rutan declared. "If you look back in history, at every point we thought we knew the limits, but they were set by our

Space Exploration Initiative directive in 1989 to land humans on Mars by 2019. Both Zubrin and Baker were then engineers with Martin Marietta Astronautics Co. (now Lockheed Martin), in Denver, Colorado.

"One problem with the 90-Day Report [a three-month NASA study following President Bush's directive] is that NASA used an irrational trajectory" from Earth to Mars, Zubrin explained. NASA intended to minimize the time away from Earth. For a round-trip, 23-month voyage, the astronauts spend 22 months in space and 1 month on Mars. "I don't think that makes any sense at all," Zubrin stated. "If minimum time in space is your figure of merit, don't leave home. Maximizing your

technology. We have explored only 3 percent of the ocean floor of planet Earth, so there is certainly much left to be explored there. I think when we are able to look down there with other than a handful of deep submersible vehicles—and make it so that individuals by the thousands can do that—you'll find that we'll learn a lot."

To date, only a few hundred have bodily made that trip. The rest of us have gone vicariously, through public educational television, perhaps. And Rutan thinks that is valid and more the way of the future. "I think we have only another 20 years or so before we won't be going to Africa to look at animals in person. We will have virtual reality resolution of all our senses just as good as our regular senses, so you could have a African experience without putting your body on the continent," he speculated. "I really think that 30 or 40 years from now, we'll explore the planets and hopefully the stars through virtual reality. We'll really be able to explore them but we won't necessarily put our physical body there.

"We have some interesting chapters coming up. And I really hope I can go out and do things that are justified based on having fun and getting fulfillment from pulling off something that is courageous, rather doing what is defined as research for some development. I don't know just exactly how the future will play out, but I have been known to do some outrageous things when I get bored," he laughed.

◆

Elbert L. "Burt" Rutan is founder and president of Scaled Composites in Mojave, California. He was elected to the National Academy of Engineering in 1989 for "leading the engineering design, construction, and testing of a series of remarkable aircraft, including Voyager, *the first aircraft to circle Earth unrefueled."*

time on the planet is why you're going there." Zubrin's Mars Direct counter-proposal lasts 30 months and could be done for less than $40 billion. But the trajectories it chooses mean that the trips each way are only 6 months, and the explorers spend 18 months on Mars where they use local resources to produce oxygen and return propellant. [For more details, see Zubrin's box "Do we really need an armada to explore Mars?" p. 204.]

But why bother going at all to a desert planet? The Mariner probes and Viking landers in the 1960s and '70s revealed that the canal-watered lush Mars envisioned by astronomer Percival Lowell and novelist Edgar Rice Burroughs was

It looks like Mars but for two features. The terrain is entirely white, and a dog waits at the end of the ramp from the lander. To built public support and awareness of its Mars Direct scheme—and to demonstrate a number of the technologies—the Mars Society is developing a Mars simulation base on Devon Island in Canada's Parry Islands, at about 75° north latitude. The base unit will mimic the plans for a Mars Direct lander and habitat, and will have a four-person crew. Fort Devon also is near the regions explored by Amundsen, the program's patron saint. © *Mike Koonce for the Mars Society*

but an arid wasteland, scoured by sandstorms. Hope rebounded in the late 1990s when meteorites found on the Antarctic ice on the Earth were traced to origins on Mars—and their insides held tantalizing (and controversial) hints that Mars harbored bacterial lifeforms. Moreover, images returned from the *Mars Global Surveyor* in orbit, and the *Pathfinder* lander and *Sojourner* rover on the surface, showed unmistakable evidence of ancient flash floods.

"We have two fundamental sets of questions," Zubrin replied. "The first set revolves around the issue: 'was there, or is there still, life on Mars?' Mars had liquid water for a billion years, longer than it took life to appear on Earth. If we find any form of life on Mars, we will have proven that the processes that lead to the rise of life on a planet are not uncommon. And that opens possibilities across the cosmos.

"Then we shift to a much more interesting and important question," Zubrin continued. "Can there—will there—be life on Mars [again]? Mars is not just an object of scientific inquiry. It is a world. It has enough water, locked up in the polar ice caps and permafrost, for an ocean 600 feet deep across the planet. It possesses as much fresh water as the Earth." The presence of so much water is one reason Zubrin is so convinced that it is viable to send humans to Mars to live off the land.

Zubrin contends that the Mars Direct approach requires no new technologies, just the integration and improvement of existing ones to develop the vehicles that would go. "In year one, an Ares rocket, derived from the Space Shuttle, sends an unmanned payload consisting of the Earth return vehicle, an SP-100 nuclear reactor, liquid hydrogen, and empty methane and liquid oxygen tanks. Once there, the SP-100 rolls out and powers the ship. It sucks in carbon dioxide [the primary ingredient in Mars' thin atmosphere] which reacts with the hydrogen, and gives you methane and liquid oxygen." The two can power internal combustion engines on surface rovers as well as the rockets for the return trip; the liquid oxygen also replenishes the crew's air supply. If ice is nearby, so much the better: it can be electrolyzed into oxygen and hydrogen.

A year later, after enough propellants have been refined for their return trip, the first four-person crew is launched. Every two years, two more rockets head for Mars, just like the first pair—one unmanned and one manned. According to Zubrin, the landers could scatter across the surface, but within range of the original rover, or they could be clustered to build an Antarctic-style research village.

Mars is not the only other place in the solar system, besides the Earth, to have water. The discovery of water and ice on three of Jupiter's largest moons—Europa, Callisto, and Ganymede—has led NASA's Jet Propulsion Laboratory to start designing a unique autonomous underwater vehicle (AUV), called *Proteus*. It is envisioned that after a lander punches its way through Europa's 2-kilometer-thick ice cover—a challenge that NASA is studying—*Proteus* would swim around the icy depths and look for signs of life.

To be sure, the waters are cold and hostile, probably liquefied only by the pressure of the overlying ice, or possibly by submerged volcanic activity or high salinity. But explorers on Earth have discovered that life thrives everywhere. *Proteus* may take its first test dive in Lake Vostok, which is the size of North America's Lake Ontario, 3.7 kilometers under the Antarctic ice. Like Europa, the reason for it being liquid is unknown: pressure, a hot spot, salt, or perhaps a combination of these.

Down to the Ocean Floor

"I love it. I think it is grand, and we should support *Proteus* with our hearts and souls and dollars," exclaimed oceanographer Sylvia A. Earle, whose own exploration led to a world-record solo dive, for a 2^1/$_2$-hour walk in a special diving

ROBERT ZUBRIN: Do we really need an armada to explore Mars?

THE MARS SOCIETY

"If you look back at the history of human exploration, it's possible for people to do it on a shoestring budget," said Robert Zubrin, a former engineer with Lockheed Martin Co. and the leading advocate for the Mars Direct concept he co-developed. In fact, Zubrin feels, the secret to success in pioneering expeditions is to keep things simple and to live off the land.

As the best example of how to do Mars right, he points to the 1903 expedition by Norwegian explorer Raold Amundsen to find a Northwest Passage—a hoped-for waterway from the Atlantic to the Pacific around the north coast of Canada.

"*Gjoa* [Amundsen's ship] was one of the great ships of human exploration—yet it was small enough that it could have fit inside the payload bay of the Space Shuttle," said Zubrin. "In many respects, Amundsen's mission was like the first human expedition to

hardsuit 385 meters (1,265 feet) under the sea. "But at the same time, not to find similar enthusiasm and make a similar commitment to explore our own oceans is literally like having your head in the clouds.

"We need easy, cost-effective working access throughout great depth," Earle continued. "We don't really have working access to most of our planet." Up to now, expeditions have consisted of "just bulldozing our way in a heavy-handed, expensive manner for a handful of people every now and then to glimpse something of the nature of this most important environment."

The lack of access "discourages ready use for those who want to go to ask

Mars might be. It lasted three years and involved six people in small, cramped quarters. Amundsen was frozen in for two years on King William Island. He and his crew got fat from eating caribou that they shot in the area. They brought dogs and sleds and they explored. They spent the summers traveling hundreds of miles. Among other things, they discovered that the Earth's magnetic poles move."

> **'If you want to go to Mars, you can't do it in 30 years. You've got to do it in 10 years or less. The longer we wait, the harder it will be.'**

More than 100 attempts to find the Northwest Passage failed before Amundsen's expedition. The Royal Navy tried no fewer than 30 times, once with steam-powered convoys. Zubrin feels that the very complexity of the earlier expeditions was the cause of their downfall. In 1845, "Sir John Franklin took two ships and 127 men. They even brought along fine china, everything that a Victorian gentleman needs," Zubrin said. "They were never seen alive again."

That is the same flaw he saw in NASA's initial response to President George Bush's 1989 call for reaching Mars by 2019 (itself a notion set by NASA).

"If you want to go to Mars, you can't do it in 30 years," Zubrin asserted. "You've got to do it in 10 years or less." He and a growing crowd of adherents in the Mars Society, with chapters around the world, are promoting a simpler plan to enable a series of four-person, 30-month expeditions that would refine air, water, and return fuel from Martian resources while crews range across the planet. "It's like a pioneer living for several weeks off the mass of a bison, just for the price of transporting a few musket balls and cartridges," he said of *in-situ* resource utilization.

---

questions that don't have an immediate economic payback," she mused. Manned and unmanned submersibles are equally expensive, she pointed out, because both require large support ships for transport and operations—ships that can cost as much as $50 million to develop and $300,000 a day to operate.

Her solution is similar in philosophy to Zubrin's: "Think small, have several variations on small manned and robotic devices, and think like the ocean." By that, she means viewing the chemical corrosiveness of sea water and the pressures at depth "not as something you fight, but rather as something you work with. Seals actually get better under pressure. Glass spheres get stronger with increas-

'Let's use the technology that's already on the shelf—and do it on a shoestring budget.'

In-situ resource utilization sharply reduces the mass that has to be launched from Earth, and eliminates the perceived need to develop a nuclear rocket. "A mission relying on advanced propulsion as a centerpiece is a loser," Zubrin declared. "I don't think those things could be ready for flight in 10 years. Let's use the technology that's already on the shelf, or near it."

Technologically speaking, "we are in a better position to send people to Mars today than we were to send them to the Moon three decades ago," Zubrin said. "There has never been a time when a nation was materially more capable of launching a great age of exploration than the United States is today. The main obstacles now are political. It is incumbent upon us to seize this time.

"We got to the Moon coming off the strength of our victories in World War II. As the memories of these accomplishments fade, though, the memories of these people become myths," he warned. "The longer we wait, the harder it will be."

♦

Robert Zubrin is the founder of Pioneer Astronautics, a space exploration and development firm in Indian Hills, Colorado. He is a former chairman of the executive committee of the National Space Society, Washington, D.C., and a co-founder of the Mars Society, based in Indian Hills, Colorado.

ing pressure. If you work with the ocean and with the pressure, you'll come up with some of the neatest and simplest breakthroughs, [such as] seawater hydraulics, where the water itself is used as hydraulic fluid. You are more likely to come away with success, rather than trying to cram our preconceived idea of systems that work in air into an environment where they are not really at home."

Technology advances are letting remotely operated vehicles (ROVs)—ones controlled by scientists at the surface—evolve into true autonomous underwater vehicles (AUVs)—ones that make decisions on their own—that can be dispatched on a mission and then recovered days or weeks later. "There is an ongoing evolu-

Water, water, everywhere, and oceans to drink. Europa became one of the most important worlds for seeking extraterrestrial life following the discovery by the *Galileo* spacecraft that this moon of Jupiter is covered with ice rafts (below) the size of San Francisco Bay. These are believed to drift atop a liquid ocean, somewhat like Earth's continental plates moving atop the molten mantle. The water is kept liquid by the pressure of the overlying

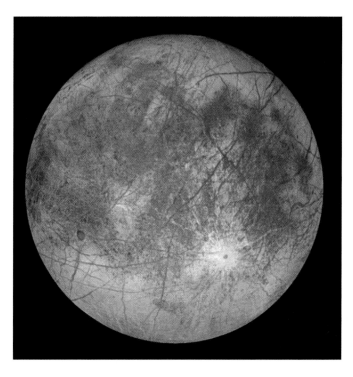

ice, and possibly by tidal forces from Jupiter stretching the moon's rocky core. Colors in the ice may be the result of minerals. *NASA and Jet Propulsion Laboratory*

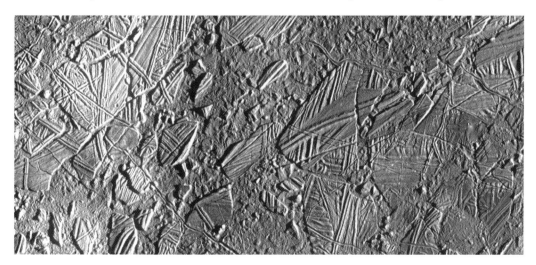

tion, from human divers and humans in submersibles to ROVs and now to AUVs and various hybrids," observed Joseph Vadus, president of Global Ocean Inc., Potomac, Maryland. "The perennial argument in selecting manned or unmanned systems depends primarily on the mission and functional needs." In Vadus' opinion, operational autonomy and reliability of AUVs have become a reality with improvements in sensor fusion—computers to integrate navigation and other inputs into a single sense of place and direction—and energy sources that have endurance for long, untended voyages. In the hybrid arena, autonomous fish-like vehicles are employed, along with a fish that is radio-controlled with an attached receiver.

JOSEPH R. VADUS: Will humans live in cities floating on the oceans?

"Long before people are living on the Moon or Mars, I feel there will be people living in floating cities," said Joseph R. Vadus, a long-time practitioner of ocean engineering, what he defines as engineering in the ocean environment.

Floating cities are a concept straight out of science fiction, yet attainable. But there's no fiction about Vadus' concept. Population pressures already are pushing humans to inhabit the surface of the seas, as Japan is doing. "Japan has long been driving the move out to sea," he said. "They have no choice but to go out to sea, either onto artificial islands or floating platforms. About 70 percent of Japan's land is largely unusable because it is mountainous, and the people are on the coastal area," he said.

In the last two decades, Japan has built several artificial islands to expand coastal communities seaward. The most prominent are Kansai International Airport in Osaka Bay, and Port Island, near Kobe, supporting more than 20,000 inhabitants plus hotels and light

ROVs and AUVs could be sent to explore for gas hydrates—methane gas (natural gas) locked in ice at great depth under the ocean. "In gas hydrates, globally there is about twice the amount of existing fossil fuel sources," Vadus explained. One possible use for ROVs or AUVs would be to "characterize their location and makeup and learn how to economically extract the methane gas." Also, gas hydrates on the sea floor in deeper water, if disturbed by pressure or temperature changes, could release gas from its structure and destabilize the sea floor, causing problems for oil production platforms working in progressively deeper water and adding to greenhouse gases. ROVs or AUVs could also help in exploring and extracting man-

industry. Kansai will be the cornerstone of the $140 billion Osaka Bay Marine Corridor Project that will link nine coastal and artificial islands by 2020.

But other designs are possible, and technical issues abound. "Would they be moored or free-floating?" Vadus asked. "There is great variation as to how mobile they could be. You have to be careful, because if you are too close to a coastline, you are going to be subjected to tidal waves and big hurricanes with storm surges."

'Long before people are living on the Moon or Mars, there will be people living in cities on artificial islands and large floating platforms.'

But if a floating city were more than 320 kilometers (200 miles) away from its homeland—outside the standard economic control zone—what would that mean politically? Could it lead to political as well as physical independence? Floating cities "could very well be floating sovereign countries," Vadus suggested. "The Law of the Sea [which establishes territorial waters

Divers tend the Automated Benthic Explorer (ABE) as it is lowered into the water from its support ship. The builders expressed their hopes by painting the nose with "NCC 1701 B"—the designation for television's starship Enterprise. The ABE will boldly go where no human has gone before, touring the deep ocean for several days at a time, untethered to a support ship. © *Al Bradley, Woods Hole Oceanographic Institution*

and economic exclusion zones] may be adjusted in time to allow that. When you are in international waters, I don't know how you can declare yourself independent. I'm not a lawyer. But the issue has been raised of floating cities declaring themselves sovereign. You erect a mailbox and declare yourself independent."

It might start with something as simple as a mailbox—look at how the economy of the Cayman Islands has developed and benefited from clever banking laws—that then allows someone to turn technical concepts into sovereign nations.

"All my life, I have been in R&D in forward areas where people would laugh at you for advanced concepts," Vadus said of his work in undersea vehicles. "Conceptually, critics would say that is not going to happen. And slowly, all these things have come to pass. It is inevitable. The burgeoning population will be increasingly dependent on the oceans for life's essentials of food, water, living space, recreation, and transport. The benefits are far more tangible than what outer space can offer in the decades ahead."

◆

Joseph R. Vadus is president of Global Ocean Inc., Potomac, Maryland. As a co-leader in the U.S.-France Cooperative Program in Oceanography, his marine technology program included a deep-ocean survey project that led to discovery of the wreck of RMS Titanic. *For over 20 years of notable service, the President of France awarded him the French Order of Merit. He has been recognized by the governments of Japan and Mexico for his services. A senior life member of IEEE, he received the IEEE Centennial Medal and Distinguished Service Award. He is a fellow in the Marine Technology Society and a Fellow in the U.K. Society for Underwater Technology. He is the author of more than 80 technical publications, and holds 10 patents (4 pending) in radar, computers, and electronic systems.*

ganese nodules (long known about but uneconomical to mine so far) and deep ocean vent lifeforms that have biotechnology and pharmaceutical potential.

Another use for ROVs and AUVs might be in finding ideal locations for the controversial dumping of wastes in deep, seismically quiet areas. "There have been proposals for ejecting low-level radioactive waste in the subduction zone" where one of the earth's tectonic plates is plunging beneath another and carried internally into the earth's crust, Vadus said. "There have been other proposals of depositing the waste in the middle of the Pacific Ocean, because in the mid-Pacific at great depth the sea floor has been stable for over a million years."

Is it Europa or Lake Vostok, Antarctica? The view may look the same in either place as Proteus, proposed remotely operated submersible, pokes around in search of possible life forms. The discovery of extremophiles—organisms living in extreme conditions on Earth—has led scientists to theorize that Europa may be our best bet for finding rudimentary life elsewhere in the solar system. Meanwhile, Lake Vostok, liquid water under a 3.7-kilometer (2.3 mile) sheet of glacial ice, may harbor hardy lifeforms marooned there centuries ago when the lake formed. While the Lake Vostok mission will be on a tether, a Europa mission will require advances in autonomy and communications. *NASA and Jet Propulsion Laboratory*

But even with highly capable ROVs and AUVs, Earle feels there will always be room for humans in exploration. "You want to have the experience, judgment, and all the other things that human beings can bring" to unfamiliar missions in unfamiliar environments, she said. "You even want the sense of adventure, the ability to communicate and articulate the experience, which is done much better by a human than a piece of metal or plastic," she continued. "No matter how good your cameras are, they can't sense and feel and tell anybody else why they love being there or why they don't love being there."

Predictions: Space and Ocean Tourism

Where explorers once trod and sometimes died, tourists ultimately follow. By the end of the twentieth century, even the mighty and forbidding Antarctic and Mount Everest became commercial tourist destinations.

Could outer space and ocean depths attract tourism in the twenty-first century?

"I think it is going to be up to the small companies to generate spacecraft pilots for space tourism, or to reach out for the exploration of the ocean floor," said Rutan.

"The year 1908, when Orville Wright went to Europe and flew his airplane in Paris, was important because that was when the world essentially found out

Two tourists head for an up-close and personal view of the sea. Cruise time is limited to the diver's safe dive time, but the tourists are fully immersed in the underwater world, rather than having to view through portholes. The acrylic dome actually gets stronger with depth since it is evenly compressed at all points. Such dives usually are limited to about 30 meters (98 feet), the sunlit zone where most oceanic life is found. *SeaMagine Inc.*

that powered flight in the atmosphere is not just possible but looked pretty easy. Within four years, there were 40,000 pilots in 39 countries.

"We found that it's possible for humans to go into space in 1961. But space research since then hasn't been done by entrepreneurs, or businessmen, or by people for profit or fun," he observed. "And now there is a sense that private individuals can't go into space on their own. Because of that, despite the work of several nations and billions of dollars, in 39 years we have only about 60 pilots and 500 passengers.

"I predict that someone is going to show that it is not just feasible but easy and safe to go into space—at least to the extent of an Alan Shepard kind of flight [*Freedom 7* in 1961], where in 15 minutes you go up to the altitude of low Earth orbit satellite, and back. Once that happens, I believe that within another four years we will have 40,000 spaceship pilots. Once you have the pilots, you will have the tourist passengers," Rutan said. "My goal is to generate the pilots."

Vadus hypothesizes that people may eventually live ever on the move on the sea. [See his box "Will humans live in cities floating on the oceans?" p. 208.] The first step is more visitors. "Underwater tourism has accelerated in the last decade with multi-viewport passenger submarines exploring coastal waters and reefs," Vadus said. Tourists can visit the wreck of the *Titanic*—4 kilometers (2.4 miles) down—on Russia's *Mir* submersibles for $32,000 a ticket.

"I think the passenger submarine idea is a great one," Earle exclaimed. "I believe the benefits are manifold, not the least of them being you are allowing a lot of people who might not be willing to put on a mask and flippers to see the

Japan's ambitious Osaka Bay Area Marine Corridor Plan, to be developed during 2001–2020, aims to produce manmade wildlife corridors and business/industrial islands. The backbone of the corridor will be a multi-tunnel system for trains, cars, trucks, and utilities. Aeration systems and sunlight collecting systems will conduct oxygen and light to the bay floor, which now is obscured by pollutants. A waterjet system will also start exchanging stagnant water with fresh sea water. Success here could lead to similar efforts farther from shore. *Research Committee on Osaka Bay Area Mega-Infrastructure*

ocean from the inside out. They'll get to meet fish swimming in the ocean, instead of just with a lemon slice and butter." In her view, that experience might give them a new perspective, "an attitudinal shift that I think is extremely important. You begin to see this whole planet in a different way." [For Earle's perspective on the necessity of such attitudinal shifts, see her box "What does it take for people to realize that technology-induced climate change is jeopardizing our very lives?" Chapter 10, p. 228.]

In Earle's opinion, tourism might even extend to the greatest depths on the Earth—12 kilometers (7 miles) under the surface. "Why not 7 miles?" she asked. "People bicycle for 7 miles. They ride horses for 7 miles. They climb mountains all by themselves under circumstances that are far more hazardous, in many instances, than the worst scenario I can think of that they might encounter in the deep sea if they're in a protective envelope that is appropriately designed. Why not?" ◇

CHAPTER 10

THE ENVIRONMENT

HUMAN ACTIVITIES ARE PERTURBING the natural balance of the atmosphere-land-ocean system that supports life on earth. Of that there is little doubt, conclude the four environmental experts interviewed for this chapter.

Yet unknown—and the subject of intense debate—are the resilience and adaptability of the earth in absorbing the load of waste and pollutants and the deprivation of habitats and species. Have we seen actual measurable evidence of global warming? Even if we have, should the amount be of concern? What are the potential consequences? And at this point, can those consequences either be ameliorated by the earth itself—or averted by other human activities?

One key reason so much uncertainty and wrangling exist over such fundamental questions is that, until recently, it has not been possible to show an unambiguous cause and effect between human activity and global change. And many effects have multiple causes, some of which are wholly natural.

But at this millennial moment, all that is changing. Scientists and engineers are working on a number of measuring and computing systems to establish environmental cause and effect on a global scale. Humans now have the scientific knowledge, the computing power, and the mathematical models to simulate the

Left: *Pathfinder*, an unmanned electrically powered aircraft, flies over Kauai, Hawaii, during a demonstration of its abilities to make long-term flights to monitor the environment. Aircraft derived from *Pathfinder*'s lessons will allow long-term monitoring of the atmosphere at a level of detail not possible with satellites. *Pathfinder* set a world altitude record for solar-powered flight of more than 15.3 kilometers. *NASA/Dryden Flight Research Center*

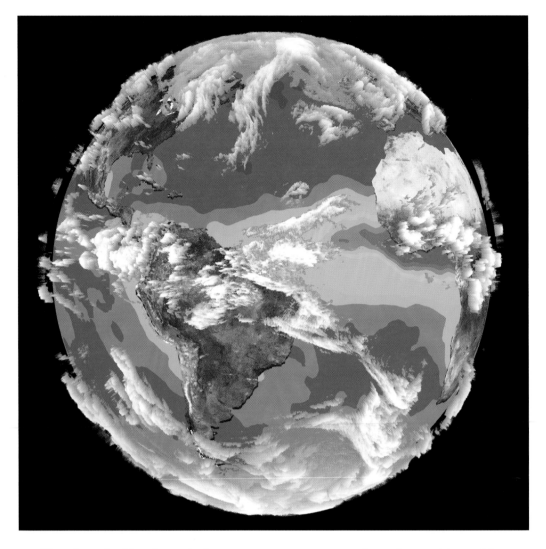

Satellites launched by the United States, Japan, Europe, and Russia provide a wide range of measurements of the atmosphere, land surface, and oceans. This image depicts a composite showing cloud data—the most familiar, thanks to TV weather forecasts— ground vegetation cover, and sea surface temperature. *NASA*

complex circulation patterns in the atmosphere and ocean that transport heat, nutrients, and pollutants around the globe. Moreover, many regional measurements of environmental change can now be compared.

Offering their viewpoints on the future of humanity's environment are four luminaries from widely differing backgrounds. Stewart Brand, whose best-selling *Whole Earth Catalog* (published annually from 1968–72) drew early and widespread public attention to global ecological issues, reminds us how powerful technological action may be in preserving or destroying our earth's life support system. Ghassem Asrar, associate administrator for earth science at the U.S. National Aeronautics and Space Administration (NASA) headquarters in Washington, D.C., suggests how researchers are blending information from satellite observations of the earth with computer models to simulate the earth's climate system and to predict future climate changes. Sylvia A. Earle, oceanographer and marine biologist who is the former chief scientist of the U.S. National Oceanic and Atmospheric Administration (NOAA), speaks to the need for a widespread monitoring of the "weather" patterns deep beneath the seas. And M. Granger Morgan, pioneering developer of university programs for examining the impact of technology on public policy, offers an engineering systems perspective for how policy makers can make wise global decisions, even in the face of uncertainties in scientific data.

Engineering with Responsibility

"A temperate climate is *not* a given," warned Stewart Brand, co-founder of the consulting firm Global Business Network in Emeryville, California. "Human engineering is now powerful enough to set into motion climate events that we can't undo.

"It is possible with increments of global warming, for example, to shut off the Gulf Stream," he continued, referring to the thermohaline circulation in the ocean that transports half of the earth's solar heat energy from the equator to the poles. In almost conveyor-belt fashion, driven by salt and heat, dense cold water

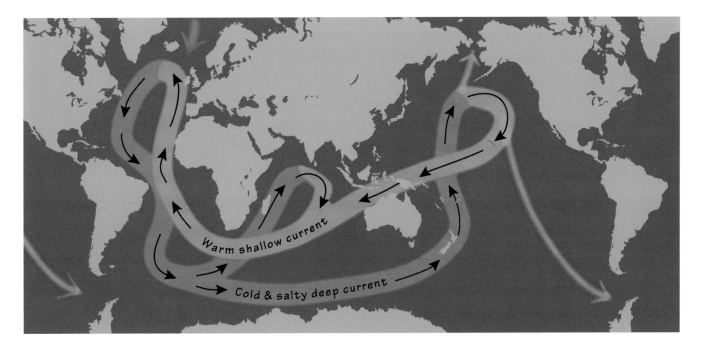

The "Ocean Conveyor Belt," showing cold, salty currents flowing deep (blue) and warmer shallow currents (red), is depicted over an image of the ocean floor based on radar altimeter measurements of sea height by the European Space Agency's *Earth Remote Sensing-1* satellite. The speed and physical conditions in this thermohaline circulation could affect global warming (or cooling) trends. *Dave Dooling*

STEWART BRAND:
Why should engineers take the long view?

"Historically, engineering has been superbly focused on content. But sometimes its practitioners have figured that context is confusing or irrelevant," reflected Stewart Brand, co-founder of the consulting and training firm Global Business Network in Emeryville, California. "But context is the next larger frame of reference in space or time or consequences.

"The attitude of 'this is for me, not for you, and you have to take it because I'm the powerful one' means it's bad engineering," said Brand. "It's elective blindness that condemns one to running down blind alleys. Uncaring engineering is bad engineering in the slightly larger picture, which—lo and behold!—we're all in!"

Illustrative of the range of past uncaring engineering, he cited "the talk of some software programmers of 'the user's a loser'" all the way up to major corporations that have compromised safety standards because of cost pressures. A notable example of the latter, he said, was Exxon Corp.'s now-infamous decision in the 1980s not to use double-hulled ships for the transportation of crude oil, resulting in a devastating oil spill from the grounded Exxon *Valdez* that despoiled hundreds of miles of Alaskan shoreline.

in the North Atlantic sinks to great depths and flows south around the Cape of Good Hope, driving warm surface water north in the form of the Gulf Stream.

But if the earth were to warm just enough for the glaciers of Greenland to begin to melt, geological evidence suggests that the influx of fresh water into the North Atlantic could shut down the flow of deep cold water, which would stop the counter-flowing Gulf Stream. "Europe freezes and can't grow food," Brand said, pointing out that the latitude of London is about the same as that of frigid Labrador, but is kept temperate only by the Gulf Stream.

Similarly, aquifers of fresh drinking water, "thanks to civil engineering, are

The Exxon *Valdez* incident, in fact, was a turning point in "making corporations realize that they *had* to take into account the environment, not just in cosmetic public-relations terms but in terms of running a sound business," Brand said. As a result, now, at the beginning of the twenty-first century, "corporations are much less the enemy [of the environment] than my leftover Left friends make out," he said. "Partly that's because their current leadership is part of a younger generation that grew up with environmental consciousness as part of their frame of reference and sense of national responsibility. And corporations have also discovered that, as a business, they can't keep their best people if those people don't feel good about working there. The employees know what a company is *really* doing—and if they're unhappy about the company's policies about the larger context, they vote with their feet and go to the competition."

Brand, through his firm Global Business Network, helps major corporations and government agencies around the world do scenario planning—thinking through alternative possible futures 10 to 20 years ahead to ascertain how the organization might be affected in each case. "A 10- to 20-year time frame is a big jump for people who are used to working from year to year or quarter to quarter," he said. "But big companies need to be aware of their larger context because they can't turn on a dime. How would their fortunes be affected if oil prices increased sharply? Or if war broke out in certain parts of the world? Or if the U.S. stock market tanked? Should they just shrug and live from quarter to quarter and hope for the best?

"If, on the other hand, they prepare for several different, quite plausible futures—three or four is plenty—then their corporate strategy will be far more robust," Brand said. "The company will have prepared itself to be stronger in its ability to handle variability without getting knocked around." Among other things, scenario planning causes company leaders to "read newspapers in a completely different way," he noted. "They give more attention to the world context and look for signs of early indicators as to which alternative future may be the scenario being played out."

under threat," Brand pointed out. "They are being emptied fast, but they refill slowly." Moreover, the world's global cod fisheries, "with the total awareness of all parties, have been fished to devastation as a result of smart new fishing technologies," such as acoustic fish finders, accurate positioning data, and satellite observations of thermal patterns that are correlated with the known preferences of fish.

Brand is not a Luddite. "Technology has generally offered more solutions than problems," he said. But he is a believer in the proverbial ounce of prevention being worth a pound of cure.

Moreover, he has found that scenario planning often causes corporations to behave with greater responsibility toward the environment, its region, and its employees. "If a company is looking at being prosperous for the next 20 years, it finds it has to start thinking about the health of the regional economy—because it won't be able to attract workers to a town nobody wants to live in. And it starts thinking about the education of local children if it wants an educated workforce in 20 years. So the next thing you know, the company is encouraging its employees to serve on local planning commissions and school boards," Brand said. In short, companies discover that "if you look farther ahead into the future than just the next quarter or the next year, things that look like altruism actually become self-interest.

"So in the training of engineers—more history, please, and more humanities," Brand suggested, with the reminder that those who are unacquainted with history are doomed to repeat its mistakes. "It's important that engineers not just build stuff, but that they build the *right* stuff. In the twenty-first century, it's imperative that they see how the content fits into the broader environmental and cultural context."

◆

Stewart Brand was the founder and editor of the original Whole Earth Catalog *(1968–72)—whose many editions sold 2.5 million copies worldwide, and for which he received a National Book Award in 1972. He also was the founding editor and publisher of the* CoEvolution Quarterly *(1974–85, which now continues as* Whole Earth Magazine*). A prolific author, his book* Two Cybernetic Frontiers *(New York: Random House, 1974), on Gregory Bateson and cutting-edge computer science, was the first to use the term "personal computer" and to report on computer hackers. His latest book is* The Clock of Long Now: Time and Responsibility *(New York: Basic Books, 1999).*

The global ocean-atmosphere system and biological communities "are clearly resilient," he acknowledged. "But we don't know where they *stop* being resilient. They're not infinitely forgiving. And we don't know where the tipping point is. We don't know which are the keystone species until they drop out of the arch. Then, when something goes bad, we can't get it back." As just one example, he cited the apparently irretrievable losses of rain forest in Latin America that were cleared to allow agriculture. The soil is very thin and low in nutrients, however, and easily washes away—and, once cleared, is not easily reclaimed by forest. "We can't get it back for a long time," Brand said. This fact is a matter of major concern

A stunning artist's concept called atmospheric panorama shows Earth as it is viewed by satellite sensors in various portions of the spectrum that are tuned to see different aspects (moisture, clouds, El Niño, etc.) against the background of hurricane Fran.
NASA/Goddard Space Flight Center

now that scientists are coming to comprehend that the rain forests house the majority of the earth's terrestrial species, and that species loss goes as a power law of area cleared.

"Engineers need to be alert to the *systemic* implications of technological change," Brand urged. Many systemic changes, he pointed out, are brought about on time scales much longer than the average corporation's focus on the next quarter's earnings or the engineer's focus on the next doubling of circuit densities on a computer chip.

"We need to manage the fast-moving parts of our civilization in a way that takes good care of the slowly moving parts," Brand said. "The fast-moving parts are technology, commerce, and now infrastructure—all engineering-intensive. The slowly moving parts are nature and culture. Culture is where we park a lot of our behavior having to do with the long term—universities, religions. We need the slowly moving parts because they are our cultural memory and constraint—they sift out the non-obvious lessons that wisdom is supposed to be about. They represent the health of the system.

"When everything is moving quickly, mistakes cascade. When things are moving slowly, mistakes instruct," Brand pointed out. [See his box, "Why should engineers take the long view?" p. 218.]

And when it comes to technological impact on the climate and the earth's capability to support life, he said, "we don't want civilization moving in a mistakes-cascade mode."

GHASSEM ASRAR: How can we best invest in the next generation of scientists and engineers?

"One challenge is how to train the next generation of earth scientists and engineers who will help realize the potential societal benefits from investments in Earth observations from space," said Ghassem Asrar, associate administrator for earth science at the National Aeronautics and Space Administration (NASA) headquarters in Washington, D.C.

In Asrar's opinion, the best way is to catch the interest of young scientists and engineers who have graduated within the last five years and get them directly involved with actual investigations. To this end, when he was the chief scientist for the Earth Observing System, he established the NASA Earth System Science Fellowship Program, which has trained more than 500 young scientists since its inception in 1991.

Now he is spearheading a program to reach doctoral students. "We are inviting them to propose to NASA to be-

NASA PHOTO BY BILL INGALLS

Powerful Eyes in the Sky

Remote-sensing and weather satellites have given humans a "God's eye" view of Earth's seas, air, and land. Between 1999 and 2003, more than 30 satellites of the multinational Earth Observing System (EOS) will be launched to give even more: an unprecedented integrated global view, available to everyone in near-real time.

The sheer variety of measurements will "help us to move from specific disciplines—such as ecology, geology, hydrology or atmospheric sciences—to looking at the whole earth as a coupled system," explained NASA's associate administrator for earth science, Ghassem Asrar. "The benefit of such interdiscipli-

come the principal investigators for small space missions that they would design, develop, launch and operate while still in graduate school," said Asrar.

This is not a one-time stunt. NASA is "going to make this a routine part of our way of doing business" in the twenty-first century, Asrar explained. "We have set in place a mechanism similar to what we use to select missions from senior scientists and engineers. Our goal is to support these students so they see their experiments completed in the three to five years that they are in graduate school."

But Asrar's ultimate ambition is to reach even younger students—undergraduates, students in two-year colleges, and even children in grades from kindergarten through high school. "Education in the broadest sense is an essential ingredient in this strategy," Asrar said. "An educated citizenry and policy-making sector is essential to our success. Therefore, we want to invest in the next generation of scientists and engineers all the way."

As part of his commitment to interest youngsters and educate the general citizenry in the potential of earth observations, Asrar was instrumental in helping establish the Earth Today gallery of the National Air and Space Museum in Washington, D.C., which opened in 1998. Real-time multiple-panel displays of images and other data is shown from various earth-observation satellites such as TOPEX/Poseidon (which measures the ocean's surface topography as an indicator of the ocean's heat content), the Total Ozone Mapping Spectrometer (which measures the concentrations of atmospheric ozone that blocks ultraviolet radiation from reaching Earth's surface), and weather satellites in polar and geostationary orbits. At this exhibit, Asrar said, visitors of all ages can "see what is happening each day to the biosphere, the atmosphere, and to land masses." Daily updates to Earth images are "completely different from the way we studied the earth in the past few decades," Asrar noted, when it was common for

Borneo burns in a computer-enhanced image produced from the Advanced Very High Resolution Radiometer on polar-orbit weather satellites. AVHRR can track the loss of world forests by the reduction in reflections unique to chlorophyll, and hot spots from burning forests. *NASA/ Goddard Space Flight Center*

'An educated citizenry and policy-making sector is essential to our success.'

weeks or even months to be required for the analysis or synthesis of a single image.

Some 20,000 people were reported visiting the Earth Today gallery during the first few weeks after its opening, Asrar said. Moreover, an estimated 20,000 people continue to pass through the gallery every month. "Now, museums in New York City and Chicago, plus the two Disney locations [Disneyland in Anaheim, California and Disney World in Orlando, Florida] have expressed interest in replicating this capability," Asrar reported with satisfaction. "This is truly the environmental information age at the service of society at large.

"The problem right now is that the volume of data is too large for a PC-based system. It requires a Silicon Graphics workstation and a special projection system. But," he added, "at the rate that technology is moving, the capability will become routinely available on the Internet in the middle of the next decade, so even schools can have access to it—the best science project possible."

◆

Ghassem Asrar has written more than 70 scientific papers, primarily in the fields of Earth surface energy and radiation exchange and biosphere/atmosphere interactions. Before joining NASA in 1989, he spent a decade in academia. He is the author and editor of several remote-sensing reference books, including Theory and Applications of Optical Remote Sensing *(New York: John Wiley & Sons, 1989) and* The State of Earth Science from Space: Past Progress, Future Prospects *(Woodbury, New York: American Institute of Physics Press, 1995).*

nary knowledge will help us refine conceptual earth-system models" so that investigators can better model "the past behavior of the earth's climate and environment as a proxy for future predictive capabilities."

Unlike earlier-generation remote-sensing data, EOS's information will be available to users almost as soon as it is obtained. "We will have a smart information node in orbit. Algorithms will process information onboard and downlink it to the desktop computer of end users and decision makers in near-real time at the cost of placing an international telephone call," Asrar said. Unlike past missions, the principal investigators—the lead scientists who developed the instruments—

will not have exclusive rights to the data. Anyone can buy a copy of the raw data simply for the cost of reproduction and shipping. "In other words, we are not trying to recover any up-front cost for developing the system. We believe that the [raw] data should not have any restriction on it," Asrar said. [One example, of freely using remote sensing data for public education, is described in Asrar's box "How can we best invest in the next generation of scientists and engineers?" p. 222.]

Most powerfully, EOS will "help us move from qualitative remote sensing to true quantitative remote sensing" through a massive effort to calibrate instruments and analytical techniques, Asrar said. "We made sure that all these instruments were calibrated and referenced to each other and [standards at] the National Institute of Standards and Technology in Gaithersburg, Maryland. The EOS program is also standardizing the various analytical methods that scientists use to turn the reflected images into a scientific conclusion. Thus, everyone knows what methodologies, principles, and assumptions the EOS investigators are using," he said. "Now, we are trying to come up with standards for data sharing, data visualization, grading of the information, and data analysis."

The Need for Eyes under the Ocean

As powerful as satellite technology can be, however, "there simply are some things you cannot measure from a platform in the sky," declared Sylvia A. Earle, founder and director of Deep Ocean Exploration and Research in Oakland, California, and explorer in residence at the National Geographic Society in Washington, D.C.

"Satellites monitor such information as temperature, salinity patterns, and patterns of plankton growth over broad areas," Earle said. But all those observations are just skin deep. "We have a decent idea of currents and chemical composition at the ocean surface. But what's below 10 feet?" Earle asked. "What's below 100 feet? What's below 1,000 feet? What's happening at the ocean's average depth of 2 1/2 miles? What's happening at the maximum depth of 7 miles?"

In her view, satellite observations "urgently need to be complemented by systems that provide 'ground truth,'" she stated.

One way to get needed information, Earle stated, is to install a network of "weather stations"—underwater observatories—in the sea. Rutgers University in New Brunswick, New Jersey, has been monitoring conditions at sites off the New Jersey shore for many years, and the National Oceanic and Atmospheric Administration and the Environmental Protection Agency are monitoring nutrient loads and toxicants from certain stations that are repeatedly visited. But to gain meaningful insight about the nature of the oceans overall, many more such stations are needed, Earle said.

"We need a network of sensing instruments positioned at key locations and depths. If you want to find a continuous picture of temperature and salinity and the changing rate of growth of certain organisms, you need a station that sits in the same location 24 hours a day, year in and year out, ticking off the information as times change."

Moreover, there is a place for undersea laboratories—another technology that has not developed to the point where some of its proponents thought it would by the turn of the millennium, she said. In her experience—which included living under water for one or two weeks at a time on nine different occasions—undersea

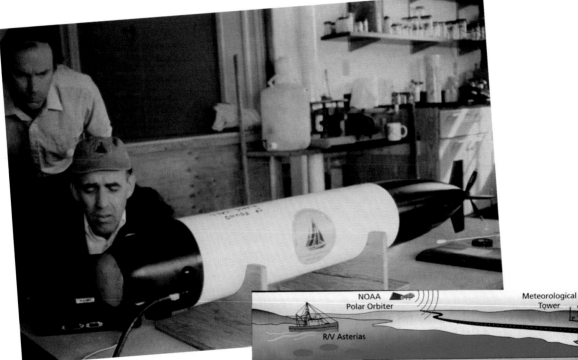

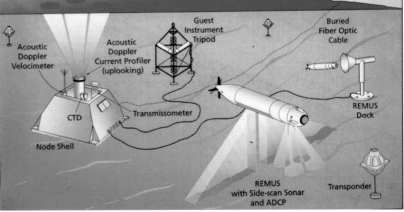

REMUS (Remote Environmental Monitoring UnitS) are low-cost, unmanned underwater vehicles designed for coastal monitoring and multiple vehicle survey operations. REMUS vehicles routinely do 20-kilometer surveys while collecting underwater data. Advanced batteries will let the vehicles range up to 80 kilometers at 2 kilometers per hour, and in excess of 100 kilometers at lower speeds. It is just 132 centimeters long with a body diameter of 29 centimeters. © *Shelley Lauzon (top) and Jayne Doucette (bottom), Woods Hole Oceanographic Institution*

laboratories give researchers "the gift of time," she said. "You actually live in the laboratory itself."

She noted with eagerness Japan's heavy investment in deep-sea technology. Among the reasons for this investment is the need to monitor earthquakes originating in the deep sea so as to be able to warn of potential earth-shaking events and plan survival. Second, "they have focused on something that few nations are: trying to understand the microbes that dwell in the deep sea, especially those around hydrothermal vents," Earle said. "They see potential medical and industrial implications.

"I hope that the twenty-first century will be a time when we record as much interest, enthusiasm and dedicated support for ocean technologies as the twentieth century did for space technology," Earle continued. "We have to make a difference. Humans have explored less than 1 percent of the deep ocean. We urgently need to understand the nature of the planet that is our life support system. For us to be content not to know is sheer folly. [For her fuller viewpoint, see her box "What does it take for people to realize that technology-induced climate change is jeopardizing our very lives?" p. 228.]

Dealing with Global Uncertainty

"Obviously, what [policy makers] would like to have is all the science nailed down before they make a policy decision. But they can't. The problems are too large and too much is at stake. They can't afford to wait [for years] until all the facts are in before making a decision," said M. Granger Morgan. Morgan is Lord Chair Professor and head of the Department of Engineering and Public Policy at Carnegie Mellon University in Pittsburgh, Pennsylvania, and co-director of the university's Center for Integrated Study of the Human Dimensions of Global Change.

"When I first started working on environmental problems 30 years ago, people did virtually no uncertainty analysis," he recounted. "Someone would produce their best estimate for exposure to an environmental hazard and their best estimate for dose response, and out would come an answer," he said. "But someone else might use the same data and different best estimates, and out would come a different answer. And depending on the point each side wanted to make, one would pick the optimistic numbers and paint a rosy picture, and the other would pick the pessimistic numbers and paint a bleak picture.

"The problem was, both would show a piece of the truth, but neither would show the whole truth," Morgan explained. In other words, there was no "probability density function, or betting odds" on the likelihood of the result. "In engineering, people had routinely dealt with uncertainty for years. But in the whole

SYLVIA A. EARLE: What does it take for people to realize that technology-induced climate change is jeopardizing our very lives?

"The oceans are our life support system. If anybody doubts that for a minute, consider what the planet would be like without the oceans," said Sylvia Earle, founder of Deep Ocean Exploration and Research and Deep Ocean Engineering Inc. in Oakland, California, and a research associate at the Smithsonian Institution in Washington, D.C.

"No matter where on the planet we are, we absolutely do the oceans' bidding," she continued. "The oceans drive the planet's weather and climate. Shifts in ocean currents [El Niño/Southern Oscillation] off the coast of Peru affects the worldwide price of corn, creates droughts in Africa and flooding in the [U.S.] Midwest. It has a profound impact on our lives, but it is fleetingly acknowledged. Then we go back to business

environmental policy area, they did almost nothing, although they faced much larger uncertainties."

Two factors in the last three decades have changed all that: the entry into environmental policy of people with engineering backgrounds familiar with uncertainty analysis, and the advent of powerful computing tools that now make it relatively straightforward to quantify uncertainty.

"The basic idea is to use a systems approach and simple techniques like conservation of mass and energy—the same amount must go in as comes out," Morgan said. Execution still can be complex, especially when it involves complex

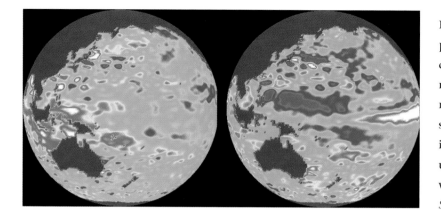

El Niño, the climate phenomenon that controls droughts and monsoons, is readily measured by changes in sea surface levels that indicate warm water upwelling (red and white). *NASA/Goddard Space Flight Center*

as usual, to the grocery store, to watching television, buying clothes. We lose sight of the context by which we all *live*."

All too often, she said, when people think of the ocean, "immediately, they think of what can be taken out of it: the exploitation of valuable goods such as seafood, oil, gas, or treasures lost at sea—things that have a clear market value today. But there is a greater importance and underlying urgency that is, in part, blinded by the attention being given to those more obvious commercial values.

"In the last century, we have initiated changes in the oceans—and in the planetary system as a whole—that are affecting the processes that make life on the earth possible," Earle continued. "We have altered the chemistry. It appears we are nudging the temperature regime toward increasing warmth. We have been cavalier about the use of natural systems and species, even though we know that we can't put them back again once they are gone. Whether it is sea turtles or whales, tuna fish or cod, we are just mining them out of the ocean as if we don't care if they are there tomorrow. We've talked a lot about sustainability and all the rest, but certainly we are not practicing what we preach.

"Meanwhile, we are generating the load of greenhouse gases. One of the important

computer models such as those used in modeling regional air pollution or assessing global change—but it's a start.

Engineering analytical methods have helped "reveal unanticipated consequences," Morgan pointed out. "Which is environmentally better—disposable or cloth diapers? Paper or ceramic plates? Electric vehicles or internal combustion engines? The answer often depends on where you draw the boundaries in the life cycle of the product. Do you include the pollution released by the power plant generating the electricity to heat the water to wash the diapers and plates? Or the lead that may leak into the environment when the half a ton of batteries are recy-

reasons for exploring the deep seas is to try and size up what is happening to the methane [natural gas, composed of an atom of carbon and four of hydrogen] that is sequestered as gas hydrates, natural ice formations in the deep ocean. Increase the earth's temperature just a little bit, reduce the ocean's pressure just a little bit, and we could trigger a massive release of methane" from the gas hydrates, Earle said. "Bubbles go off into the atmosphere, increasing the greenhouse effect, which could further increase the warmth and accelerate the process.

"And," she added, "the amount of methane stored in the [hydrate] formations in the deep sea is stupendous. It is believed to exceed all other hydrocarbons [fossil fuels] in whatever other form they exist."

Some investigators have suggested that a mechanism like a massive release of methane from the oceans might have created past cataclysmic changes in the earth's climate, Earle noted. "We don't know what causes ice ages to come and go. All we know is that they are a fact of Earth's history. And it appears that, instead of the process being extremely gradual all the time, there are occasions that it happens very swiftly—perhaps within the span of a human lifetime.

"The extent to which we modify, change, alter the natural system has profound consequences to our future—to our *immediate* future, to the lifetime most of us can look

The Monterey Bay Aquarium Research Institute's remotely operated vehicle (ROV) *Ventana* completed dive number 1,500 on Saturday, October 10, 1998, off Santa Barbara, California. This accomplishment represents more than 7,300 hours that *Ventana* has explored the deep sea. *Ventana* is equipped with lights, a broadcast-quality video camera, sensors, sampling tools, and specialized equipment for various types of deep-sea research and public outreach. *Monterey Bay Aquarium Research Institute*

forward to having, not just to the future of our children and beyond. We are heading towards situations that may not be to our liking. And, personally, I am not as optimistic as some of my colleagues about finding what they call 'quick fix' solutions," she said.

"We can use some of our technology to remind us every single day of our lives that we are here only because the oceans exist," Earle concluded. Two technologies in particular "are pretty simple." One, she said, "is called the two-by-four. It is that stick of wood that smart farmers in the [U.S.] South use to get the attention of mules. They whack the mule between the ears and all at once the mule responds. We need to do that to ourselves, and say, 'Wait a minute. Wake up. There is an ocean out there and that is where our future is. The oceans are in trouble. And if they are in trouble, so are we. Get busy.'

"The other piece of technology is that simple thing called a mirror," she continued. "Everyone ought to pick one up and look hard in it. Don't wait for the next guy to take action. What is stopping *us* from using our good brains and the extraordinary opportunities? *We* have to make a difference."

◆

Sylvia A. Earle is explorer in residence at the National Geographic Society in Washington, D.C., and head of the National Geographic Society's 1998–2003 Sustainable Seas Expeditions. She has dived deeper than any other human—1,000 meters—in an apparatus that was part submarine and part dive suit, a record that has remained unbroken since 1986. An expert on marine algae, Earle's research concerns the ecology of ocean ecosystems and the development of technology for access to and study of the deep sea. The author of more than 125 scientific, technical and popular publications, her most recent books are Sea Change: A Message of the Oceans *(New York: G.P. Putnam Sons, 1995) and* Wild Ocean *(Washington, D.C.: National Geographic Society, 1999).*

cled?" To better quantify problems, environmental engineers, collaborating with economists, have begun to use input/output tables that describe economic flows in the economy, coupling them with data on energy use and the release of toxic materials. And that has helped them devise "green" design techniques that better inform the development and disposal of products.

Such approaches "have transformed the ways in which environmental policy analysts and regulators view hazards," he said.

"Policy makers now know there is no way to escape uncertainty. Instead, they must weigh possibilities and make judgments in the face of uncertainty in a sys-

Man's impact on his own habitats is easily seen in aerial thermal infrared images of Sacramento, California (top), Baton Rouge, Louisiana (middle), and Salt lake City, Utah (bottom). Red spots are the hottest and usually correspond to roads, parking lots, and other concrete or asphalt structures that reradiate visible sunlight as heat. Green and blue areas correspond to cooler areas such as rivers and to sections where the moisture in trees absorbs and retains heat. Already, Salt Lake City is using these data to mitigate sources of ozone pollution before the 2002 Winter Olympics. *NASA/Marshall Space Flight Center & Global Hydrology and Climate Center*

tematic way. Rather than trying to get a single number, they now try to draw a probability distribution—a distribution of odds on the various numbers—so that the bets placed on behalf of society are more informed. Done well, uncertainty analysis breaks the deadlock by showing the whole picture rather than adversarially developed partial pictures." [For Morgan's perspective on a particular environmental issue, see his box "Why is it urgent *now* to investigate low-carbon sources of energy?" p. 234.]

Predictions: Helium 3 Fusion

"By 2050, some observers project we'll have run out of all economically recoverable fossil fuels and the places to put the residues from fission reactors," declared Wilson Greatbach. Although most widely known for his invention of the implantable cardiac pacemaker, for which he has been inducted into the Na-

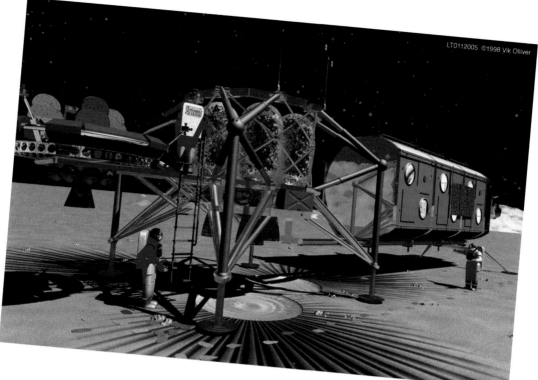

LT0112005 ©1998 Vik Olliver

In this artist's concept, astronauts prepare
to explore the lunar soil and set up an automated mining operation that would
bring back helium 3 that has been cooked into the lunar surface by the solar wind.
Helium 3 would fuse more readily with deuterium (heavy hydrogen) in a number of
fusion energy schemes now being studied. Even when fusion technologies are worked
out, helium 3 is extremely rare on Earth, but could be "mined" from the lunar soil.

© Vik Oliver, The Artemis Society (top), Dave Dooling (bottom)

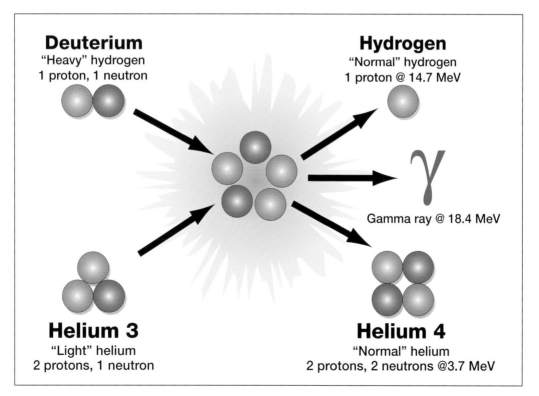

Deuterium
"Heavy" hydrogen
1 proton, 1 neutron

Hydrogen
"Normal" hydrogen
1 proton @ 14.7 MeV

γ
Gamma ray @ 18.4 MeV

Helium 3
"Light" helium
2 protons, 1 neutron

Helium 4
"Normal" helium
2 protons, 2 neutrons @3.7 MeV

M. GRANGER MORGAN: Why is it urgent *now* to investigate low-carbon sources of energy?

"Most of the lay public does not know that carbon dioxide—the principal 'greenhouse gas' implicated in global warming—originates from the burning of fossil fuels," declared M. Granger Morgan, referring to the results of formal surveys of public opinion about climate change conducted in the mid-1990s. Morgan is head of the Department of Engineering and Public Policy at Carnegie Mellon University in Pittsburgh, Pennsylvania, and co-director of the university's Center for Integrated Study of the Human Dimensions of Global Change.

"If they don't understand that carbon dioxide results from coal and oil combustion as a result of demand for electrical energy, then they can't understand that energy conservation, along with renewable energy and other low- or no-carbon energy sources, are the key to dealing with the problem. The facts are simple, but they need to

---

tional Inventors Hall of Fame in Akron, Ohio, one of Greatbatch's current research interests is various sources of alternative energy for humans on the earth.

In Greatbatch's opinion, all alternative sources such as biomass, geothermal, tidal, solar, and wind will meet only a quarter of the projected demand for electricity by 2050, when the world is estimated to be supporting 10 billion people. In his personal view, that leaves only one long-term viable source of energy: thermonuclear fusion. [For differing opinions of other technology experts who favor nuclear fission instead, see the boxes by John G. Kassakian on p. 34 in Chapter 2 and Robert A. Bell on p. 58 in Chapter 3.]

be understood" by both the public and by lawmakers, Morgan continued. Otherwise, the result is poor or inadequate public policy and regulation.

"The heat balance of the earth is well understood," Morgan explained. "The amount of radiant energy the earth receives from the Sun has to exactly balance the amount of heat the earth re-radiates into space—otherwise the earth heats up and fries. Now, most of the solar energy is received at optical wavelengths [the wavelengths of visible light]. And the atmosphere is transparent at optical wavelengths," which is, of course, why we can see through it to view the Sun, Moon, and stars.

"But once the solar energy is absorbed by the clouds, the land, and the oceans, it must be re-radiated at much longer infrared wavelengths," heat wavelengths that are longer than optical wavelengths, Morgan continued. "The atmosphere is not transparent to infrared, because infrared is absorbed by water vapor and by carbon dioxide. And the more carbon dioxide there is in the atmosphere, the more opaque the atmosphere becomes to infrared. So the heat cannot escape and the equilibrium temperature of the earth will rise. This is the 'greenhouse effect,'" he noted, so named because the carbon dioxide acts

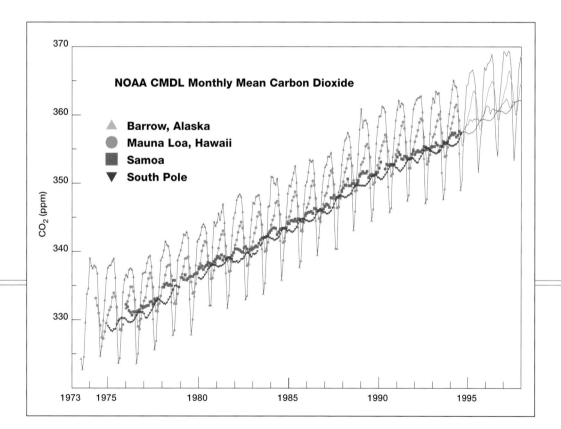

Carbon dioxide levels in Earth's atmosphere have been rising readily since scientists starting recording levels from one end of the planet to the other. The sine-wave variation may be due to seasonal trends associated with plant growth and human activities, while the upward trend is generally attributed to industrial activity. *NOAA Climate Model and Diagnostic Laboratory*

> **'There's no question what happens when an atmosphere has a lot of carbon dioxide—on Venus, the surface temperature is high enough to melt lead.'**

somewhat like the transparent glass of a horticultural greenhouse that traps heat from the sun.

In and of itself, the greenhouse effect is certainly not bad. "We have it to thank for the existence of life on Earth," Morgan pointed out.

But what could be threatening to that very life is the potential for a runaway greenhouse effect, should the atmosphere become too laden with carbon dioxide.

"There's no question what happens when an atmosphere has a lot of carbon dioxide," Morgan said. On the neighboring planet Venus, which is about the size of Earth and where the atmosphere is 95 percent carbon dioxide, "the surface temperature is high enough to melt lead," he said. "The question no one can answer is how high we can jack the amount of carbon dioxide in Earth's atmosphere before there are significant [temperature and climate] effects."

What alarms environmental scientists is that "carbon dioxide has a long lifetime once it's in the atmosphere—about a century," Morgan explained. As a result, even if all carbon dioxide production from fossil fuels could be stopped right this instant, much of the carbon dioxide produced up to now would remain in the atmosphere for more than a hundred years before being absorbed by the oceans and plants. Even more worrisome, not only has carbon dioxide production not stopped—it's accelerating as the developing countries step up their own combustion of coal to generate electricity for their populations.

"There is so much momentum in the system that we almost certainly will *double* the

---

"There are three approaches to fusion research," Greatbatch explained. "Two of them rely on fusing atoms of hydrogen into helium, as happens in the sun." As there is no bottle on Earth that can hold reactions of such high temperatures and pressures, the first approach—as typified by the Toroidal Fusion Test Reactor (TFTR) tokamak at Princeton University in New Jersey and the Joint European Torus (JET) in the Culham Laboratory in Oxford, England—uses magnetic confinement. Although the approach works, no reactions have yet been sustained for longer than several seconds—still impractical for the commercial production of electricity.

The second hydrogen fusion approach uses inertial confinement to ignite the

amount of carbon dioxide in the atmosphere by the end of the twenty-first century," Morgan stated, "and the amount rises exponentially from there. We can argue about how rapidly or how much, but there's no question it's just a matter of 'when.'"

Because much of the future production of carbon dioxide is anticipated to come from developing nations, some politicians contend that the developed countries should not curtail their own production of carbon dioxide until China and India agree to curtail theirs, Morgan noted. "But if you divide the current amount of carbon dioxide produced by a country by the number of people, we [in the United States and Europe] produce 20 to 50 times the amount per person than the developing nations. And for many decades, most of the carbon dioxide in the atmosphere will have been put there by the developed world. So, for the foreseeable future, if anyone has the responsibility to do any 'heavy lifting' in low-carbon energy technologies as well as in other strategies for reducing emissions, it must be the developed world," he declared.

"We *don't* know how far we can push the ocean-atmosphere system before we hit one of these nonlinearities" such as a major change in ocean circulation or the loss of polar ice, Morgan brooded. The only way out he sees is through energy conservation and cost-effective low- or no-carbon energy systems. "But now, just at the time we really need to build intellectual capital that will allow the market to create such systems, we have a major research shortfall. There is a serious under-investment today in research in basic energy technology such as photovoltaic materials for cost-effective solar cells, membrane materials for better fuel cells, and wide band-gap semiconductors for power electronics."

What alarms him further is the attitude in many developed nations that free markets, not governments, should take the lead in spearheading such work. "The market didn't drive computer science at the outset," he pointed out. "DARPA [the U.S. Defense Advanced Research Projects Agency] built it and then the market took over. But in en-

fusion reaction. A spherical pellet containing hydrogen is placed at the concentric focus of multiple beams from high-power lasers. The beams compress the fuel so fast that it will react long enough to produce net energy before it blows apart. In theory, "a pulse of energy is given off, which creates heat to make steam to turn a turbine to create electric power," Greatbatch said. "A fraction of a second later, another capsule comes down and another pulse is generated," and so it goes. Although this approach has been long explored by the Laboratory for Laser Energetics at the University of Rochester, New York, and at the Lawrence Livermore National Laboratory in California, "it has yet to achieve net energy generation,"

'Today there is serious under-investment in research in basic low-carbon energy technology.'

ergy technology today, there is no dynamic like that. In fact, it's the reverse—the Department of Defense in its quest for dual-use technologies is looking to ride on the market's coattails. As a result, investments in basic energy technology research is modest and not growing rapidly." Furthermore, with the deregulation in U.S. electric power utilities, he said, the utilities' funding to the Electric Power Research Institute for renewable energy resources has been slashed by a third and given a much shorter term and applied focus.

Reducing carbon emissions later in the twenty-first century "will be a whole lot more expensive if we have to do it without having laid the groundwork that will allow us to avoid expensive mistakes," Morgan warned. "We need to start taking this issue seriously *now*."

M. Granger Morgan, who received his IEEE Fellow award in 1988 "for leadership and pioneering contributions to research and teaching in applying engineering to the areas of technology and the public policy," is the co-author (with Max Henrion) of Uncertainty: A Guide to Dealing with Uncertainty in Quantitative Risk and Policy Analysis *(Cambridge University Press, 1990).*

he said. "But that's where government money is going, because funds have been cut for the American TFTR and the European JET devices."

There is, however, a third approach to thermonuclear fusion—inertial electrostatic confinement—that is far more efficient, Greatbatch said. Instead of using hydrogen, this process can fuse two atoms of helium 3—the lighter isotope of helium—into one atom of the common isotope helium 4. In the reaction, two protons are released with the energy of half a million electron volts, he said. Still in the laboratory—primarily at the Fusion Technology Institute of the University of Wisconsin at Madison, led by Gerald Kulcinski—the process has already achieved reactions by fusing helium 3 with deuterium (a heavy isotope of hydrogen).

Because of challenges posed by basic physics, "practical implementation probably will need to take place in several phases," said Greatbatch. First will come deuterium-tritium, deuterium-deuterium, and deuterium-helium 3 reactions. "Later in the century may come the ultimate helium 3-helium 3 reaction," which has the potential of generating the most electrical energy, he said.

In Greatbatch's opinion, helium 3 fusion is the "perfect" process. "The fuel is nonradioactive, the process is nonradioactive, and the helium 4 residue is nonradioactive—the stuff used in kids' balloons." Compared with hydrogen fusion, the helium-helium fusion reaction may be easier to induce and to confine: "Commercial electrostatic-confinement power plants would need only a 1,000-pound spherical cage in a high-vacuum enclosure in contrast to the 250-ton supercooled electromagnet in the tokamak," said Greatbatch. And with helium 3 fusion, "electric utility power stations will be pollution-free," he said, as they will no longer be discharging the carbon dioxide that contributes to global warming.

There's one big problem: there's almost no helium 3 on Earth, aside from a minute percentage in natural gas and in the decay products from the tritium in nuclear bombs.

But, Greatbatch pointed out optimistically, helium 3 is being thrown off all the time by the sun through the solar wind. Because it is in an ionized (charged) form, when it hits the earth's magnetic field it is diverted away from the earth. But it has been landing on the moon for four billion years, implanting itself on the surface, particularly in an iron-titanium-oxide ore called ilmenite.

"There is more energy on the surface of the moon in helium 3 than there is in all the fossil fuels on the earth," declared Greatbatch (a perspective that may differ from that of geologists trying to ascertain the amount of methane in natural-gas ice hydrates deep in the ocean. [See the box by Sylvia Earle, page 228.]). A station on the moon to mine and treat ilmenite "could produce titanium dioxide, carbon, nitrogen, water, oxygen, and other chemicals useful in supporting a manned lunar colony," he said—an option that is actually being explored by the U.S. National Aeronautics and Space Administration. Liquefied helium 3 could be transported back to the earth. The energy content of helium 3 is so great that "a single space shuttle load of 25 tons would power all the current electrical needs of the United States for a year," Greatbatch said. Moreover, "it has been calculated that the cost of establishing the lunar colony, the shuttle, and the mines, if amortized over several decades, would give us energy at a cost that is less than we now pay for a barrel of crude oil."

But, he added, if there is to be any chance of helium 3 fusion becoming a practical reality by 2050, "we had better start *now*!" ◇

CHAPTER 11

WAR AND PEACE

"IT IS BECOMING A MORE DANGEROUS WORLD rather than a less dangerous world," cautioned Norman R. Augustine, a respected former aerospace executive who spent much of his career helping the United States develop weapons to prevent war or, if that failed, win it. The Western allies won the Cold War, but now, along with their former adversaries, must gird for new kinds of war as technology becomes available to anyone with a grudge. "In many respects, the loss of the nuclear deterrent has made the world safe—safe for conventional war or unconventional war," he added.

Offering their views on what may become "conventional" in the future are three experts known for their work in developing arms and a fourth made famous for efforts to halt their development in irresponsible hands. Augustine is popularly known for his witty series of Augustine's Laws which explain why things—especially in the aerospace and defense world—don't work as common sense would dictate. He now is a professor at Princeton University's School of Engineering and Applied Science. Richard L. Garwin, Senior Fellow for Science and Technology at the Council on Foreign Relations, is a pioneer in U.S. satellite intelligence systems.

Left: The USS *Decatur* (DDG 73) fires two Standard Missile-2 anti-aircraft missiles in a test off Hawaii. The Navy proposes using a variant for theater missile defense around North Korea and in the Mediterranean Sea. The *Decatur*, one of the Navy's newest class of destroyers, has sloping sides to reduce radar signatures and a four-face phased array radar system, the AN/SPS-1B, to track targets (one face can be seen at the left edge of the picture). *Ensign Garrett Kasper, U.S. Navy*

For your eyes only: These images taken by a U.S. KH-12 spy satellite in the early 1980s (and leaked to the media) show a Soviet aircraft carrier under construction at Sevastopol on the Crimean Sea. The striping is caused by the film being developed onboard the satellite and then scanned in strips for transmission to the ground.

U.S. Government, courtesy Federation of American Scientists

David A. Kay sprang to international prominence in 1991 when he led international teams working in Iraq to declaw Saddam Hussein's weapons of mass destruction. He now is director of The Center for Counterterrorism and Technology at Science Applications in McLean, Virginia. Myron Kayton, an independent consultant and owner of Kayton Engineering Corp. in Santa Monica, California, has been involved in a number of missile and space projects since the 1960s.

Welcome to the 24/7 War

"At the beginning of the century, we basically had a two-dimensional battlefield on which military forces fought six months of the year," Augustine continued. Modern warfare has provided mobility and visibility virtually around the clock. "That revolution will continue," Augustine continued. "We will have a three-dimensional, 24-hour-a-day, 12-months-a-year battlefield made up of a global surveillance system that tracks where everything is, all the time, even down to individuals. Within that global information system, there will be an imbedded ordnance delivery system that can very quickly and very precisely deliver ordnance with very few collateral casualties."

The capabilities of spy satellites have become the stuff of legend, partly because most of what they could do is classified. The most popular notion is that America's "Keyhole" (KH-series) and "Big Bird" spy satellites can read the license plates of a car in Red Square. That may be close to the truth. Space Imaging Inc. of Boulder, Colorado, builders of the commercial Ikonos mapping satellite, promise 1-meter (3.3-foot) resolution from an altitude of 677 kilometers (423 miles).

The United States and Russia have enjoyed an immense lead in space-based reconnaissance. Commercial systems like Ikonos may end that monopoly, said David Kay: "It helps the argument when everyone has access to KH-quality data, it is harder to argue that 'Nothing is happening, it is your imagination.' Already, during the Kosovo crisis, the Serbs are arguing, 'There is no outflux from Kosovo, we are not burning villages,' when any European can call up data on the Internet and get images of it." Even weather satellites reveal smoke plumes from burning villages. During the so-called ethnic cleansing in Kosovo in early 1999, the U.S. Government started releasing Advanced KH-11 images showing what appear to be mass graves for executed ethnic Albanians. Other problems are more subtle. [See Kay's thoughts on "How much privacy will we trade for safety?" on p. 254.]

"A big problem is countries carrying out things that they don't want us to know about. They know the orbital parameters of our imaging satellites and they conduct their outdoor activities at times when they think we are not there," said Richard Garwin, who played a major role in the development of Corona, the first U.S. spy satellite. "Furthermore, a lot of things are being put underground to protect them against attack and to hide them." Even the discovery of Hussein's nuclear weapons program was a result of a tip-off by an informant after the ground war

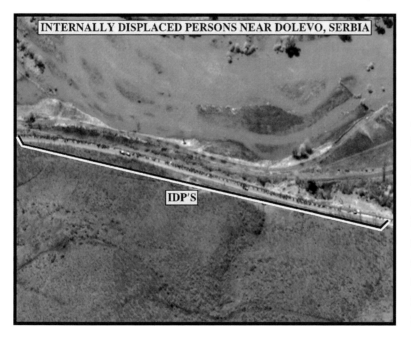

INTERNALLY DISPLACED PERSONS NEAR DOLEVO, SERBIA

IDP'S

A U.S. spy satellite photo shows internally displaced persons (IDPs) lined up on the road in Kosovo during Serbian "ethnic cleansing" in 1999. The images have their resolution degraded slightly so the satellite's full capability is not disclosed. Other images show mass graves following reports of executions.
U.S. Department of Defense

NORMAN R. AUGUSTINE: How can we watch out for a weapon that hasn't been invented?

"I think the most important development is the one not thought of yet," cautioned Norman R. Augustine, former CEO and chairman of the board of aerospace giant Lockheed Martin Corp. "If you go back to the beginning of this century and compare the situation then to the current situation, so many of the things that have been really important were ones that no one foresaw, even shortly before they become realities. That's likely to be the case in the next century, so the best thing to do is stand by for surprises."

While weapons will become smarter, smaller, and faster, Augustine does not yet see the soldier going on the dole.

"There will always be the role of will and of courage. I would be the last person to predict that the next war will be won by the side with the last computer standing. It's

ended. "Intelligence is a very important aspect," Garwin continued. "But so is having the right environment so that people know we are interested that what's going on is wrong, that most of the people in their country would believe it was wrong if they knew about it." [And that leads to Garwin's discussion of "Can we mount an effective defense without having to shoot?" on p. 248.]

Festung America

The Strategic Defense Initiative (nicknamed "Star Wars") spent about a trillion dollars and fielded no weapons. Yet its legacy lives in dozens of continuing

just not likely. The human is still likely to be terribly important. To control the ground you must occupy the ground.

"Technology has lead to major unforeseen breakthroughs throughout history, which have had the consequence of changing the character of history as well as of warfare."

'The best thing to do is stand by for surprises.'

One profound change is the spread of affordable technology. Sixty years ago the very word radar was classified. Today, baseball coaches use it to clock pitchers' fast balls.

"Most of the technologies where the leading edge is found are no longer in the military sphere but in the commercial sphere," Augustine continued. "And a consequence is that they will be available to a lot of people. If you happen to be a relatively wealthy nation, you would like to have a high entry threshold for new technologies, just as if you have an established business, you would like to have a high entry threshold in the marketplace. Unfortunately, that's no longer the case in the defense context. An individual can be a major threat to a tank with an anti-tank missile. They can be a threat to a close-support aircraft with a shoulder-fired anti-aircraft missile. They can be a threat to a capital ship with a relatively inexpensive anti-ship missile. Thus, smaller and poorer countries can be major threats to the capital assets of larger, wealthier countries. Worse yet, terrorist groups or terrorist nations that can obtain nuclear, chemical, or biological weapons could be a major threat to civilization. My bottom line is that it is becoming a more dangerous world rather than a less dangerous world."

Unfortunately, it also means that arms development and spending will not go away.

"You have to stay ahead of everybody else by just running faster than they do. That means having countermeasures to deal with things an opponent might have. That's been

projects that were given new life by Iraq's hail of ballistic missiles derived from the USSR's Scud (itself derived from Nazi Germany's V-2 missile) in the Gulf War. "Theater [short-range] defense is essential because every Third World country you're going to fight is going to be able to buy North Korean Scud equivalents," Myron Kayton said. Like the Russian-designed AK-47 assault rifle for guerrilla bands, derivatives of the Scud are becoming the great equalizers for developing nations, especially India, Iran, Iraq, North Korea, and Pakistan.

Defense could be an international venture. "Russia and the United States have a lot in common," Augustine said. "We might want to build together a ballistic

'The most important technologies in the next century are the ones we haven't thought of yet.'

the character of warfare since a caveman first threw a rock, a weapon having the disadvantage that it was reusable by the enemy. You just have to run to stay ahead.

"Think about the advent of the stirrup, and the consequence that had, and then the longbow, or gunpowder, or the machine gun, or the tank and the airplane, ballistic missiles and nuclear weapons, and stealth, and so on down the line—each a major discontinuity in combat. A nation that wishes to survive dares not miss those discontinuities; it must be willing to think out of the box to find them. And it takes money to do it. If a nation does those things, it can govern the course of combat. If it misses them it becomes the victim. That seems to have been true since the beginning of warfare. But there is no reason to think that in the next century there will not be periodical breakthroughs which no one thought of in advance. The most important technologies in the next century are the ones we haven't thought of yet."

Norman R. Augustine was elected to receive an IEEE Fellow Award in 1989 "for technical leadership and contributions to space and military defense." He received the National Medal of Technology and the National Space Club's Goddard Trophy. He is best known for Augustine's Laws *(printed in four languages) (Washington: AIAA, 1997 [6th ed.]) and is the co-author of* The Defense Revolution: Strategy for the Brave New World *(San Francisco: ICS Press, 1990). He also holds a copyright on a special calculator for baseball managers.*

missile defense system and it would be thin enough that it wouldn't undermine the basic deterrent that both nations maintain, but it would be strong enough that it would keep out the irrational rogue."

Boost-phase interceptors—such as a proposed upgrade of the U.S. Navy's ship-launched Standard Missile-2—could provide adequate defense where the interceptors could be based within hundreds or even thousands of kilometers of the launch site, as is the case with North Korea firing north towards the United States. "It's so easy it is embarrassing," Garwin said, "because the rocket flame has for decades been seen by infrared satellites high in geostationary orbit." As in Desert Storm,

U.S. forces play the "bad guys" as they set up a Scud missile site at Roswell, New Mexico, on April 26, 1997, during Exercise Roving Sands '97. Scuds, used by India, Iran, Iraq, North Korea, and Pakistan, pose a widespread threat to regional stability. To counter it, Western military forces are developing tactical anti-missile capabilities.

In a different role-playing game, a surplus U.S. Lance missile (right) played the role of a Scud missile for the infrared tracker atop a Navy missile being tested as an interceptor. *U.S. Department of Defense photo by Petty Officer 1st Class Stephen Batiz, U.S. Navy (top), U.S. Navy (bottom)*

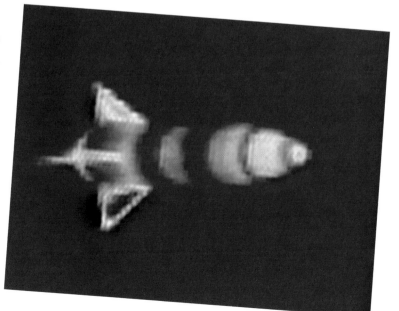

RICHARD L. GARWIN: Can we mount an effective defense without having to shoot?

"You want to dissuade people from behaving in an inimical fashion," said Richard L. Garwin, IBM Fellow Emeritus at IBM's Thomas Watson Research Center in Yorktown Heights, New York. Garwin has been involved in dozens of committees that determined the shape of U.S. weapons and strategies. "The economy can be used politically, but an economic lever might be used to keep people from being so aggrieved that they would take action against another nation."

Keeping technical capabilities as visible as possible is the best defense, Garwin argues.

"One approach is to have countries publish within their borders—for instance, in *Science* and *Nature*—the details of the countries' obligations under the Nuclear Non-Proliferation Treaty or Biological Weapons Convention so that individuals within those nations will know what is illegal for them to do as individuals," he explained. "Some of these treaties require the states party to the treaty to pass legislation to criminalize individual activity that is illegal under the treaty. More of that should be done."

an interceptor can be cued by early warning satellites. Interceptors could be made even smaller, he continued, "little things that weigh 20 kilograms (44 pounds) or so which will spread over the landscape maybe in rivers and be commanded by radio to take off and pursue one of these things" in the first seconds of flight.

Total, impregnable defense for a large nation like the United States will remain impossible, though, argued Garwin. "People talk about the necessity to protect all 50 states with a defense against ballistic missiles," Garwin said, "but that doesn't defend all against the destructive threat of nuclear weapons, hardly defends against ballistic missiles. Less than $6/10$ of 1 percent of the people live in Hawaii and Alaska,

Terrorism, attacks on diplomatic personnel and embassies, genocide, hijacking aircraft, and piracy are recognized as universal crimes. Garwin believes that designing and building weapons of mass destruction—already crimes in the United States and most nations—should be elevated to the same status. "Then there would be a basis for any country seizing any person who is involved in such activities whether or not that country's nationals were involved or it was done on their territory," Garwin says.

However, that requires actions by responsible, or semi-responsible governments. Yet, many governments are lax and look the other way for a bribe, or simply lack the resources even to control things within their own interior, he observed.

"When most of the governments really do end up signing these treaties, they ought to pay more attention to their disclosure of what they have, internal laws or not," Garwin continued. "Universal criminalization is a separate treaty. Most of the nations will sign that too. If they do, then it gives other countries the right to operate within their borders and to seize people. Right now if people work within this country—where either they are sheltered or there's a *laissez faire* attitude and they stay there—well, it's pretty hard to get them out." Witness the 10-year effort it took to extradite from Libya the two men accused of the Pan Am bombing over Lockerbie, Scotland.

"If they can never travel, then that reduces the threat quite a lot," Garwin stated. "It is not a perfect solution, but it is a lot better than nothing."

And one could also argue that if the govern-

Role-playing victims run under the water spray from a fire truck as part of a decontamination procedure following exposure to a simulated chemical agent during Exercise Cloudy Office held at the Pentagon, May 30, 1998. It involved more than 500 participants from the Office of the Secretary of Defense, FBI, Arlington County Emergency Medical Services, Metropolitan Medical Strike Team, the Pentagon's TriCare Health Clinic and special Defense Protective Service units. The purpose of Cloudy Office was to exercise force protection responsibilities and to test the response by civilian emergency components to a hostage and barricade situation complicated by the threat of weapons.
U.S. Department of Defense photo by Staff Sgt. Reneé Sitler, U.S. Air Force

ment is looking the other way that by default they have become a participant.

"Under those circumstances, those individuals in the government themselves are liable," Garwin agreed. "And they do travel. So intelligence agencies would have a field day on that" if the legal mechanisms are in place so the officials could be arrested and prosecuted.

Garwin also noted that while globalized trade has made attacks by small nations and groups easier, one aspect of that trade may make it easier to track criminal warriors: "The Internet, e-mail, for that matter, exchange of 'cookies' [electronic visa stamps that some web sites leave in your browser software] and similar things, could provide anonymous, secure reporting of forbidden activities."

'You want to dissuade people from behaving in an inimical fashion.'

Richard L. Garwin, born in 1928 in Cleveland, Ohio, received his B.S. in Physics from Case Institute of Technology, also in Cleveland, in 1947, and a Ph.D. in Physics from the University of Chicago in 1949. He was a member of the President's Science Advisory Committee (1962–65, 1969–72), and of the Defense Science Board (1966–69). In 1978, he was elected to the U.S. National Academy of Engineering for "contributions applying the latest scientific discoveries to innovative practical engineering applications contributing to national security and economic growth." He received the 1996 Enrico Fermi Award. He is co-author of many books, among them Nuclear Weapons and World Politics *(New York: McGraw Hill, 1977).*

which add greatly to the problem of a 50-state defense. So if you are doing it progressively, you wouldn't defend Alaska until late in the game." Atop that, the equation changed with the fall of the USSR. Whereas the United States faced the potential for 150 million or more deaths and destruction of all major cities, a rogue would have only one or a few nuclear weapons that he would be more likely to deliver by truck, ship, or even airliner. And that poses a different detection problem.

"If some country can threaten to kill 100,000 people or whatever—that is not something wished or tolerated—we ought keep it in perspective," Garwin said. "It's not as if each person in the country is going to be in among those relative few, one part in 2,000, who would be destroyed even if the other side could kill 100,000 people."

Chemical and Biological Warfare

Coupled with the spread of low-cost missiles is an increased willingness in some quarters to use chemical or biological weapons, even on the civil front as demonstrated by the murderous Aun Shin Rikyo cult's Sarin nerve gas attack on a Tokyo subway on March 20, 1995. Biological and chemical protection often has much in common: both must detect an invisible agent, shield troops while letting them fight, decontaminate the scene, and treat the victims. Key differences are that chemical agents work within minutes or seconds, and biological agents incubate for hours or days while an infected individual spreads them.

"We're going to see a lot of detection schemes, sniffers, maybe bugs that evolve biologically that grow in gas and turn green so that you know that the bug has discovered a gas environment," Kayton suggested. [See more of Kayton's thoughts on "How do we reduce the body count?" on p. 260.] "You would have a liquid solution and expose it to air and if the right gas gets in there the bugs react appropriately." That is, once they are released. Weapons sealed in heavy plastic or foil will be tougher to detect at ports of entry. Indeed, it would be almost impossible to distinguish quickly the different strains of *E. coli* bacteria—some are common in the human gut, some are genetically altered to make insulin, and some have killed people who ate contaminated burgers. Even so, defense against biowar agents may be possible, Garwin argues: "Passive defenses against biological attack are not so costly. Positive pressure in buildings with filtered air so no biological agent can get in through the air. It doesn't take much to reduce the hazards from an infectious agent by a factor of 100 or more. And even for contagious agents, if you can reduce the reproduction factor below 1, then you have a different threat than if you have a wildfire epidemic."

The best defense might be the oldest. "I think we ought to resume universal vaccinations against smallpox. It is really an ideal strategic biowar agent because it has been refined over the millennia to be very effective," Garwin said. At this writing, the United States leadership was debating whether to destroy its last known samples of live smallpox (the Russians are expected to do the same). Some people believe it should be saved in case it is needed to help fight a new strain that might be natural or manmade.

Guns for Everyone

One of the lessons from the aftermath of the Gulf War was that almost any-

A Russian-built, Kilo-class submarine, one of several purchased by Iran in the 1990s, is towed by a support vessel in the central Mediterranean Sea. Diesel-electric submarines like the Kilo are a serious naval challenge. They are exceptionally quiet and, when operated in narrow waters like the Persian Gulf, are almost impossible to detect because of the noise from surface traffic. *U.S. Navy*

one with enough determination and money can develop weapons of mass destruction, and that no corner of technology can be regarded as safe or obsolete. Saddam Hussein, for example, expended vast resources to refine fissile uranium using a method—calutrons—that the United States had discarded as too expensive in the early 1940s.

Guerrilla wars in the last half century were fueled in part by arms discarded by larger nations in the wake of World War II. That established a pattern that continues. As the superpowers moved to larger more sophisticated weapons, their client states continued to acquire older weapons. Shoulder-fired anti-aircraft missiles left over from the Afghan civil war are available on the black market. The United States now struggles to dissuade North Korea and China from selling variants of the Soviet-built Scud ballistic missile. Other wars and trade deals have spread enough weapons that "smaller countries can be major threats to the capital assets of larger countries," Augustine said. And if they can't buy them, they can manufacture a handful with commercially available parts from such outlets as Radio Shack or Carolina Biological Supply.

As a result, large military organizations will have to grapple with the dual challenge of keeping systems for half a century while incorporating the latest technologies. "The average item of military equipment today stays in the inventory—with the current budgets in the U.S.—for at least 50 years," Augustine observed. "And we have a technological half-life, depending on the technology, anywhere from 3 to 15 years. We have a fundamental mismatch that poses a great management challenge. This suggests that in the case of expensive items, we are going to have to find a way to extend their lifetimes. One way you do that is having platforms you keep for a long time and then continuing to upgrade those components that have very short half-lives." [On p. 244, Augustine answers, "How can we watch out for a weapon that hasn't been invented?"]

Another attractive solution is using commercial off-the-shelf components, although this can be fraught with problems. In 1998, the U.S. Navy dedicated a late-model guided-missile cruiser and a destroyer as technology test beds, including centralized computer operations based on Windows 95. One notorious test gave new meaning to "blue screen of death" when the system crashed and the cruiser had to be towed back to port. "Sometimes, a commercial off-the-shelf product is not such a good idea," Kayton said. "Windows was not intended as a real-time environment." Using off-the-shelf hardware is not unrealistic, he added, and cited the case of the commercial PowerPC 750 chip that has been "radiation hardened" under license by Honeywell for the space and nuclear environments. "Starting all over again is a kind of military thinking left over from the days of infinite budgets," he continued, "where you say I'm going to start with a clean sheet of paper and do better than everyone in the past; usually you don't do any better but you spend a lot of money."

The End of the Aircraft Carrier and the Manned Jet?

The widespread use of cruise missiles and advances in robotics raises the old question of whether the aircraft carrier, the centerpiece of U.S. Navy doctrine since 1941, and the manned aircraft, the centerpiece of any air force, are obsolete. "I don't think that manned aircraft have much of a future as fighter-bombers," Gar-

DAVID A. KAY:
How much privacy will we trade for safety?

"There is going to be tradeoff of our civil liberties and our individual right of privacy against a desire for security," said David A. Kay, best known to TV viewers for his work as the chief arms inspector in Iraq after the Gulf War. As director of The Center for Counterterrorism and Technology at Science Applications in McLean, Virginia, he worries about tracking security threats by terrorists who will bring America the sense of vulnerability that much of Europe, Japan, and other nations have had for years.

"The nuclear arms race between the two superpowers became something that was a fact of life, not a daily concern," Kay said. "That is likely to change as we find ourselves more vulnerable to people who are less defined—national groups, or even the horror of Littleton, Colorado, of individual loners that can do a tremendous harm. You are likely to see an attempt to develop defensive technologies to protect you as well as offensive technologies to punish those who are responsible." The most effective deterrent is the sure

win has argued since the 1960s, when guided air-to-ground missiles like Bullpup came on the scene. He envisions a system in which command-and-control and radar systems—airborne for an elevated line of sight into enemy territory—direct unmanned missiles carrying TV cameras. "We see these pictures at the last moment, have somebody there who for each of the missiles had been prepared with a photograph of the target from the aspect of what the missile was going to see on its way in, and the controller would simply put a finger, or a mouse, or a pencil on the place in that photograph where the missile was supposed to strike." That, basically, is what the Navy did with two standoff strike missiles in the Gulf War.

and certain knowledge that you will be found before you can carry out an attack. But that may require encroaching even farther on the very privacy that Americans cherish.

"How far that swings either way is not going to be determined by a theoretical academic balance," Kay continued, "it's going to be determined by the frequency of outrageous, threatening events. If the attacks appear to be more lethal, more frequent, more threatening, and the government seems to be less able to prevent them, then you will push further on that swing and decrease privacy or civil liberties."

Great Britain pressed ubiquitous traffic cameras into the search for Irish Republican Army terrorists. Many Americans would object to having so many eyes watch their every move. Yet, "most Americans don't realize the coverage of ATM and other cameras that are around. In some areas they are pretty dense. New imaging grabbing technology and enhancing technologies will go well beyond recording the face of someone withdrawing money." And virtually everyone has a picture on file with the state for their driver's license or passport, and often with an employer for an ID badge.

"There's no need to know who the terrorist is," Kay said. "You can do it for everyone and just pull out whom you want to know was where and, if you live in an urban region, track pretty well one's daily movements if you cross-reference it with ATM records, telephone records, particularly cell telephones which give geographic coordinates. Increasingly, you're giving up more information—fingerprinting, DNA samples, basic biometric data—to get that same license." Video systems soon will be trained to look for people who act suspicious.

"There are signs and tipoffs of hostile behavior. People don't just appear out of nowhere," Kay continued, "If you're worried about certain hostile behavior in the neighborhood of a bank or an air-

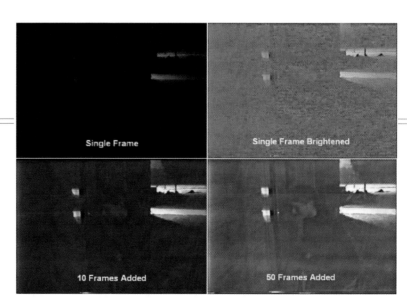

Single Frame   Single Frame Brightened

10 Frames Added   50 Frames Added

The grainy video from surveillance cameras will become clearer with a number of technologies, including NASA's VISAR, a technique developed to stabilize space launch images. Night conceals a burglar, but her identity is revealed as almost two seconds worth of video frames are composited and enhanced. Similar methods will be used by security agencies to spot terrorists in likely target areas.
Bob Moder for NASA/Marshall Space Flight Center

> **'Whether a society will let these new capabilities go ahead is . . . an issue of whether that society feels threatened enough to sign away its freedom. . . .'**

port, you worry about what are the classic body movement signs that indicate you're contemplating hostile action, things like loitering on a street corner for a set time and then going to an opposite street corner."

At the same time, increased surveillance can provoke public mistrust of the government. "This is the typical anarchist's argument for violence," Kay said. "'If we do violence, it will provoke the government into taking action that will alienate their base and we will be able to take over power.'" So modern societies will have to walk a fine line between being conquered by terrorists, and conquered by themselves. "Whether a society will let these new capabilities go ahead is not a technical issue," Kay said. "It's an issue of whether that society feels threatened enough to sign away its freedom in that regard."

◆

David A. Kay is the former United Nations Chief Nuclear Weapons Inspector, leading numerous inspections into Iraq after the Gulf War to determine Iraqi nuclear weapons production capability. He is recipient of the International Atomic Energy Agency's Distinguished Service Award and the U.S. Secretary of State's Commendation. He is director of The Center for Counterterrorism and Technology at Science Applications in McLean, Virginia.

One missile blew a hole in the side of a power plant, and the second flew through the hole to destroy the machinery. Garwin argues that smaller vessels, similar to the canceled arsenal ship and the sea control ship, will replace the 85,000-ton floating behemoths that the Navy deploys, and that more precise cruise missiles will replace the need for fighters which, after all, fly cover for attack aircraft.

"Cruise missiles are going to replace the aircraft carrier," Garwin argued. "We are proud of our aircraft carriers. It is amazing that you can do all those things that aircraft carriers can do, but in fact you don't need an aircraft carrier—you need a place from which you can launch a one-way missile. They can be cruise

Aircraft launch from the deck of USS *Enterprise* (CVN 65) on their way to strike targets in Iraq during the first phases of Operation Desert Fox, December 17, 1998. Such sights may vanish by mid-twenty-first century if the aircraft carrier is replaced by the Arsenal Ship, a latter-day monitor carrying cruise missiles and other unmanned weapons.
U.S. Navy

missiles, they can be ballistic missiles equipped with GPS. There's absolutely no reason whatsoever that you can't have a GPS terminally guided ballistic missile with all of the other capabilities." The Navy would seem to partially agree as it experiments with new designs—many of them smaller (and less expensive) than the 85,000-ton behemoths now deployed at sea—and incorporates stealth technologies. Aircraft, too, will evolve to operate with smart onboard systems, rather than pilots, under the direction of ground operators. Augustine has a different viewpoint: "Reports of the death of the manned fighter are exaggerated. People have predicted the end of the tank since before the horse, but it has not been the case. The role of aircraft will change a lot, their designs will change a lot. But I think the prediction of their extinction is premature."

Unmanned Combat Aerial Vehicles (UCAVs) drop bombs and then head back to a small-scale aircraft carrier that launched them. The UCAVs are intended to be reused, but can also be sent on more hazardous missions than would be possible when pilots' lives are at risk. *Lockheed Martin Corp.*

Predictions

The future of warfare may hold a return to gunboat diplomacy, this time based in space. Both Kay and Augustine foresee space-based weapons as one of the big sticks carried by the United States, possibly in partnership with other nations.

"Suppose you had a global surveillance system which keeps track of nearly everything all the time," Augustine said, "and you had highly reliable non-nuclear ballistic missiles that in 20 minutes could, with confidence, home in and destroy anything, anywhere on the world." Given the willingness to use such force, a missile could be dropped on a target as a warning to a foreign power to stop whatever it has been doing. The next step up, both Augustine and Kay speculate, is space-based weapons. The U.S. Air Force is developing technologies and designs for a Space-Based Laser to destroy ballistic missiles as they exit the dense lower layers of the atmosphere. Ironically, it would be "eye safe" because its wavelength would be absorbed by atmospheric moisture. But other wavelengths, even into the micro-wave range, could "zap down 20 feet from you and blow up the tree that was standing there. That sort of capability could be very sobering," Augustine said.

Kay argues that such a system is necessary in order to stop leaders such as Yugoslav President Slobodan Milosovic because the slow ramp up of a bombing campaign lets the people accommodate the danger and discomfort. "The alternative is for us to jump to the high ground, in which you base both defensive and offensive systems in space," Kay said. "That is much harder for anyone else to match or to counter, at least theoretically. It has the capability of operating deci-

It can happen here. A gunner protects civilians (Marine role players) at Camp Lejeune, North Carolina, on January 20, 1998, during an Urban Warrior exercise. As a reminder of how pervasive urban warfare may become, this is Mayberry Drive, ironically named for the fictional, idealized town in a highly popular U.S. television series (*The Andy Griffith Show*) in the 1960s. Urban Warrior is the U.S. Marine Corps Warfighting Laboratory's series of limited objective experiments examining new urban tactics and experimental technologies. *U.S. Department of Defense photo by Staff Sgt. David J. Ferrier, U.S. Marine Corps*

MYRON KAYTON: How do we reduce the body count?

"We see again and again that no-body wants to send troops anywhere because we watch them being killed on TV," observed Myron Kayton, a defense consultant. "That's a new phenomenon in history. Nobody ever knew what soldiers did except they came back marching down the street in their fancy uniforms."

Witness the immense coverage given to a single downed pilot in Yugoslavia in 1996. The solution increasingly will be machines that are out front rather than people.

"The risk has to be moved from humans to machines," Kayton continued. Eventually, micromachines wedded to microelectronics will provide miniature craft that will be the "point man" as infantry probes ahead.

"Microvehicles—small, lunch-plate-size vehicles—are going to be collecting intelligence in cities so people don't have to expose themselves in urban warfare areas. Micro-

sively against someone almost instantaneously and without regard to weather." A bolt from a space "gunboat" would be only a few minutes away.

A lot of this presupposes that the owner is the "good guy." Other governments might prefer that it be managed by the United Nations (U.N.) or some other theoretically dispassionate international body rather than the U.S. government. Kay suggested that implementation may require the multiple ownership that U.S. President Ronald Reagan proposed when he first outlined the Strategic Defense Initiative in 1984. "We may try to strike a compact with countries that says, for example, missile defense technology will shut down all missiles to an area," Kay explained.

machining gives us these unmanned aerial vehicles on a salad plate. You want something really small like these micromachined bug-size airplanes that you see being developed now. I think it is a great idea, because those things can do things that will protect humans from risk. One of them can fly down a ventilator duct or just stick its nose out to peek around the corner in an urban environment. You would have to be a pretty good skeet shooter to hit one."

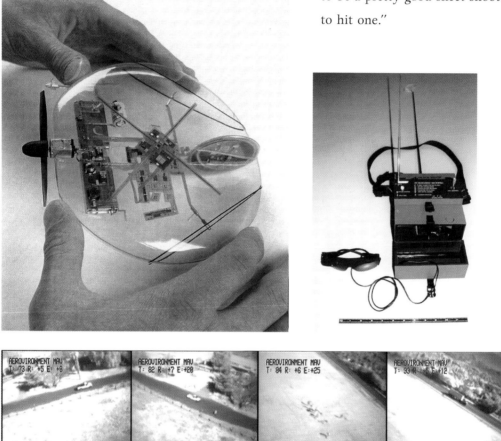

If you think of insects as unmanned aerial vehicles, you're headed in the right direction. AeroVironment Inc. of Simi Valley, California, is developing micro air vehicles (MAVs) that would provide short-range reconnaissance for soldiers in the field. AeroVironment's Black Widow—about the same size as the development model shown here—is carried in and launched from a portable case that includes the radio controls and video display (in the headset). The 60-gram (2-ounce) craft flies at 40 kilometers per hour (25 miles per hour) and is inaudible beyond 30.5 meters (100 feet), and can fly for up to 22 minutes over a 1 kilometer (0.6-mile) range. © *AeroVironment*

'Training becomes the most urgent part of our war, especially in the era when everybody has equal weapons.'

Reducing casualties on the other side will become a priority, too, although the simplest non-lethal weapon raises hackles.

"I think that is an important consideration that we may have to change our moral attitude toward gas," Kayton said. "I think that there will be research on finding chemicals that will disable somebody without long-term cancers or damage to eyes, all that stuff. The thing that got mustard gas such a bad reputation in World War I is that it destroyed mucus membranes, it literally burned out their lungs. You need to find chemicals that are not like that."

Peeking at what other nations are doing also may help reduce the need to use lethal force.

"We're going to see the return of covert intelligence," Kayton said. "It has been several decades since we shifted from covert to satellite—what they call technical means of intelligence. But the best way to find weapons is to know in advance when someone is using one. In order to counter the weapon issue, you are going to see the return of covert intelligence."

Still, satellites will remain an invaluable tool for tipping the other guy's hand.

"If [Iraqi president] Saddam Hussein or [Libyan political leader] Muammar al-Qaddafi gets some weapons, he can't train on them because we'd notice. He has to unleash his war with wholly untested stuff. So training becomes the most urgent part of our war, especially in the era when everybody has equal weapons. There are systems that you just have to drill over and over again. With complex command and control systems you've

"The problem with that is that everyone knows that means all missiles except the ones we decide to authorize. Because when you operate the defensive system it is hard to imagine how to make it not porous to your own systems if you decide to turn it off. I think that there will be part of that strategy to lessen the political opposition by agreeing to norms and management procedures. It is easier in the abstract to describe these systems than everyone has confidence in operating."

"It is not going to be universally viewed as always a good thing. In some cases people may well want to use the offensive capability when we don't want to use it," Kay noted. NATO presents a largely unified front in its actions in Kosovo, but

just got to drill and make sure everybody knows how to use them. Murphy's law is correct: things don't work right the first time, you have to keep training and practicing. I don't think that you could have Third World people train intensively without being noticed. Sigint [signals intelligence] will draw attention to training sites and invite visual observation of the sites."

> **'Murphy's law is right, things don't work right the first time, you have to keep training and practicing.'**

Ultimately, one nation's decision to act will be predicated upon an opponent's willingness—real or perceived—to react with deadly force. Saddam Hussein's buildup was detected by satellite and other means in the weeks before he invaded Kuwait in 1989, but he did not expect the West to react.

"That is part of the body bag question," Kayton said. "Remember everyone said when Hitler crossed the Rhine River into France . . . that if the French had pushed him back across the border it would have discredited him and World War II would not have happened. But no one wanted to challenge him."

◆

Myron Kayton received his IEEE Fellow award in 1981 "for contributions to the design of avionic systems for spacecraft and aircraft." He has worked on a wide range of aerospace projects, including the Apollo Lunar Module guidance system. He is the owner of Kayton Engineering in Santa Monica, California.

support and opposition varied across the globe. "Kosovo is interesting partially because the Serbians' behavior is so outrageous and so public that it is hard to imagine anyone defending what they have done," Kay said. "Other situations are likely to be somewhat grayer. Some people will argue if you have a capability like that, you should use it to prevent action. Others will say, 'No, it really is not worth it.'" As with the weapons available today, the question becomes who holds the keys, and how many keys have to be turned in series to run the engine of war. "You don't want to select one country as the universal policeman, judge, legislature for the world," Kay concluded. ◇

CHAPTER 12

PREPARING ENGINEERS FOR TOMORROW

TODAY'S ENGINEERING EDUCATION does "a good job of providing students with theoretical background in specialties, and showing them how to formulate problems so they can be solved," said Edward Alton Parrish, president of Worcester Polytechnic Institute (WPI) in Worcester, Massachusetts. "But we leave unanswered the question of whether something *should* be done, even if it *can* be done."

And that, he feels, must change for the twenty-first century.

Parrish's viewpoint—that engineering education must expose students to the societal impact of technology—was echoed by all five interviewees for this chapter, as well as by many of the luminaries whose words have appeared in the preceding 11 chapters. Moreover, all five shared uncanny consensus about three other twenty-first-century issues: the necessity of encouraging students as young as grade-schoolers in science and engineering, the necessity of transforming undergraduate engineering education from emphasis on individual competitiveness (class rank) to collaborative teamwork, and the necessity of realizing that the Internet will revolutionize university education.

All five interviewees have distinguished themselves as engineering educa-

Left: Engineering students from Worcester Polytechnic Institute work on a water quality project in Thailand. Getting students out of the classroom and into the real world gives them valuable experiences, not the least of which is understanding the long-term, real-world consequences of design and management decisions. *Worcester Polytechnic Institute*

tors. Eleanor Baum was the first female dean of a U.S. engineering school. Donald Christiansen was editor and publisher of the Institute of Electrical and Electronics Engineers' flagship magazine *IEEE Spectrum* for 22 years, which under his leadership captured four National Magazine Awards (the magazine equivalent of the Pulitzer Prize) for its detailed investigations of the societal impacts of technology. WPI's president Parrish was one of the architects of the EC 2000 educational criteria of the Accreditation Board for Engineering and Technology (ABET), revolutionizing the standards for university engineering programs in the twenty-first century. Donna Shirley was the manager of the Mars Exploration Program at the Jet Propulsion Laboratory (JPL) of the U.S. National Aeronautics and Space Administration (NASA) in Pasadena, California, and the original leader of the team that designed and built the *Sojourner* rover that was successfully landed on Mars by the Pathfinder mission in 1997. And John B. Slaughter was the first director of the National Science Foundation to have a degree in engineering rather than in the pure sciences.

Reengineering Engineering Education

"Not so long ago I would tell groups of wanna-be electrical engineering students that while a consumer or other user need only view a piece of electrical or electronic equipment as a black box, what was in the black box was the province of the electrical engineer," said Donald Christiansen, president of the consulting firm Informatica in Huntington, New York, and editor and author of McGraw-Hill's classic *Electronics Engineers' Handbook*. "I would emphasize the need of the electrical engineer to understand materials and how they were used in fabricating components.

"Now I am not so sure. With electronic systems being designed at a higher level of abstraction and with computer simulation, students may not find the need or interest in knowing what's inside the black box," Christiansen mused. "Electrical engineering as we know it may have disappeared in a decade or so. Degree accreditation rules are becoming more flexible, giving students more latitude at the undergraduate level than they've historically had, and, perhaps, making the choice of school even more important."

The need for revisions to standards had been widely felt for more than a decade, recounted WPI's president Edward Alton Parrish. By the 1980s, ABET's list of criteria by which engineering programs were evaluated had ballooned from the original $1\frac{1}{4}$ pages in the 1950s to some 20 pages. They had become so prescriptive, in fact, that "we were counting books in the library and the number of faculty, without regard to whether students were learning what they needed," Par-

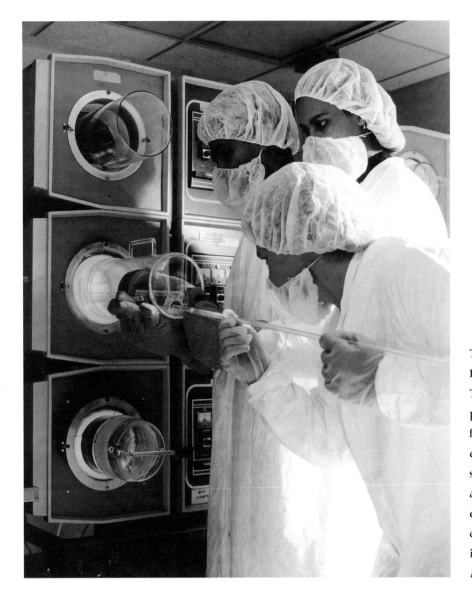

These students at the Florida Institute of Technology are preparing silicon wafers for high temperature diffusion. Engaging students in hands-on activities early in their engineering education is crucial to retaining them in programs. *Florida Institute of Technology*

rish recounted. "We had created a giant cookie cutter stamping out homogeneous engineering programs producing homogeneous engineering graduates, with perhaps some differences in quality."

Simultaneously, there was increasing criticism from industry that the young engineers so stamped out were, indeed, ill-equipped with the skills they needed in the world of real work. They were not as good as desired at working collaboratively in teams, for example, or in communicating effectively with co-workers and managers. Nor could they move gracefully from technical work into management later in their careers. Nor were they perceptive about evaluating the impact of prospective technologies on society and the environment. In short, companies complained that "too much on-the-job training was required before they could use the new hires," Parrish said.

Industry's perception was corroborated by numerous quantitative studies

EDWARD ALTON PARRISH: How can students *experience* the impact of engineering on society?

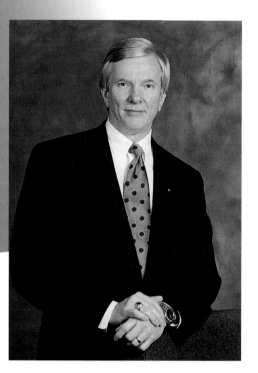

"I don't think many engineering programs are doing enough to expose students to the impact of technology on society," said Edward Alton Parrish, President of Worcester Polytechnic Institute (WPI) in Worcester, Massachusetts. "We need to make students sensitive about the impact of engineering on society—*even if* they don't have to meet environmental or other codes for their type of project.

"And just raising social concerns in a few classes is not enough," Parrish warned. "For students to really 'get it,' they need to do interdisciplinary problem-solving to develop an *implicit* knowledge base."

In his view, traditional engineering programs concentrate on what he calls explicit knowledge—"the kinds of things you get out of a handbook," he explained, "such as formulas, design rules, ranges of reasonable parameters, and even safety and environ-

conducted by government agencies and professional societies, including the National Science Foundation (NSF), the American Society of Engineering Education (ASEE), and the Engineering Education Board of the National Research Council. In addition, public resistance to the upward-spiraling costs of higher education was increasing.

All these factors led in the 1990s to "a lot of pressure in the United States for assessment and accountability" within engineering education programs, Parrish said. As a result, ABET decided to reexamine the criteria for accrediting engineering programs in colleges and universities. "The unanimous recommendation was

mental codes. These are things that are easy to put in textbooks and to teach and learn.

"Implicit knowledge, however, is not something you just read and memorize. It's something you have to experience to learn," Parrish continued. "Implicit knowledge is fuzzy compared to the square root of 2—it's almost a sixth sense a professional develops after long experience. That's why when someone on the other side of the continent is trying to design

> **'Engineering programs need to introduce social issues, especially ethics—and they need to do it *early*.'**

a complex system and gets into trouble, a company doesn't just have an expert write a memo; instead, they send the *person* to transfer the knowledge, because there are gray areas requiring experience and judgment. A lot of the appreciation of societal implications falls into the realm of implicit knowledge."

The most logical time to tie in societal issues with engineering knowledge is the capstone engineering project of a curriculum, such as a senior thesis, Parrish noted. But there's one major flaw with that approach: it's too little too late.

Because developing implicit knowledge requires so much time, "engineering programs need to introduce societal issues, especially ethics—and they need to do it *early*," Parrish said. "These are 18- to 22-year-olds. Many haven't yet given a lot of thought to societal issues. Ninety-nine percent of them just want to know in black-and-white 'how do I use this instrument?' or 'how do I use these data?' or 'how do I get the *right* answer?' But they won't know how to cope with gray areas until they face something *real*."

At WPI—which is a technological university—every undergraduate is required to do three major projects, generally in teams with faculty advisors. One of these is an interdisciplinary project requiring the student to work during the entire junior year (or for

to take a revolutionary, not evolutionary, approach—to start over completely," Parrish said.

The new ABET educational criteria—called EC 2000—will be mandatory for all U.S. engineering programs beginning in the academic year 2001–02.

"The idea was to break the cookie cutter," asserted Parrish. The new criteria not only allow, but actually "require each institution to establish its own mission and program objectives, and to set up appropriate program curricula to achieve those objectives, subject to a 'floor' of minimum criteria. Above that floor, each institution is free to do what it wants in terms of number of faculty, hours of specific courses,

'[Students] won't know how to cope with gray areas requiring experience and judgment until they face something *real*.'

two months full-time) on a real-life problem at the intersection of technology and society. Often the project is pursued in some other part of the world, Parrish said, "such as assessing environmental issues in the slums of Bangkok, or deterring the ravages of pollution on the canals of Venice, or exploring ways that technology can improve life for the handicapped in London." WPI's program of such hands-on experiential-based learning started out in the early 1970s with considerable controversy among the faculty, but now has the support of alumni and faculty alike.

"I know of no other institution that pays that kind of attention to the societal impact of engineering," Parrish said broodingly. "And it's still not enough."

◆

Edward Alton Parrish, who received his IEEE Fellow Award in 1986 "for leadership in engineering education and contributions to microprocessor-based pictorial pattern recognition," is best known for his seminal work in pattern-recognition and image processing. He recently served as chair of the Engineering Accreditation Commission of the Accreditation Board for Engineering and Technology (ABET), teams of experts who evaluate engineering programs for continued accreditation. Currently, he is involved in ABET's international activities.

and the like." Furthermore, the EC 2000 criteria specifically emphasize the need for more opportunities for multidisciplinary teamwork and hands-on experience.

All this freedom, however, is scarcely a free-for-all. As the proof of a recipe is in the eating, the proof of any engineering program is in the performance of its students. The intent of the new EC 2000 criteria is to "focus on outcome instead of on process," Parrish said. And the outcomes are specified. Engineering programs must demonstrate that their graduates have the ability, among other things, to design and conduct experiments as well as to analyze and interpret data; to design a system, component, or process to meet desired needs; to communicate

Carol Fassbinder, 18, of Elgin, Iowa, shows a model of the molecule for a new natural control she developed for a deadly parasite currently affecting the honeybee population and her family's bee-keeping business. She was one of the top winners in May 1999 at the 50th annual International Science and Engineering Fair, now sponsored by Intel. While fairs help highlight and boost motivated students, programs must reach those who lack resources and motivation.

Camera 1 photo via Feature Photo Service

effectively; and to understand the impact of engineering solutions in a global and societal context. "In the past, institutions were not required to assess what students learn," Parrish said. But the EC 2000 criteria now require evaluation of what students learn against the institution's objectives. "So there is a continuous improvement feedback loop for adjusting a program."

Aside from responding to the criticisms of industry, government agencies, and public pressure, there were several other motivations behind ABET's revision of the accreditation criteria.

One was to allow local engineering institutions the freedom to "set up programs that can meet the needs of their region's stakeholders and constituents," such as local employers, Parrish said. That aim recognizes the late-twentieth-century demographic shift of young engineers away from large multinational corporations to smaller local companies with fewer than 500 employees.

Another was to "look at engineering as the new liberal arts education of the twenty-first century, so the graduates are well versed in fundamentals and can keep up with technical disciplines, but are also prepared for success in life," Parrish continued. "Education is learning how to *learn*, as opposed to learning how to *do*. If you do a lot of training—teaching how to do a specific task—instead of educating, you do the student a disservice. Education positions engineers to keep learning."

Last, one of the most important motivations behind the new accreditation criteria was to reduce the number of mandatory technical courses in the engineering curriculum. Thus, engineering students now have more flexibility to take elec-

DONALD CHRISTIANSEN: Engineering ethics— who cares?

"Engineering, unlike science, is a risky business," declared Donald Christiansen, editor emeritus and former publisher of *IEEE Spectrum* magazine (1971–93) and editor of McGraw Hill's classic *Electronics Engineers' Handbook*. "We design and build things. Directly or indirectly, our products affect the public. And sometimes they fail.

"What the general public does not fully recognize is the extent to which engineering is a profession of tradeoffs and compromise," he continued. "Every compromise carries with it the seed of discontent on someone's part—management perhaps, or the customer. And to a lesser extent each compromise or design tradeoff carries with it the seed of an ethical lapse.

"Unlike scientists—who may be called to task for falsifying data, misleading their colleagues, or taking credit for others' work, often to obtain research funding—engineers are more likely to be deemed unethical, or at least poor engineers, for putting on the market a product that is unsafe or harms the public in some way. If we as engineers know what and where the failure mechanisms are, and we have not designed around them (through redundancy, for example) or have not determined the risk or consequences of failure to be minimal, we may be open to the accusation that we have made an unethical design deci-

tives, including liberal arts courses so they can learn the communications and management skills demanded by industry, and more humanities courses so as to "better understand the social context in which they will be practicing engineering," Parrish said. [For Parrish's further views on the subject, see his box "How can students *experience* the impact of engineering on society?" p. 268.]

Broadening the Young Engineer

Christiansen thoroughly approves of the trend in encouraging engineering students to understand the societal context of their technological work. He espe-

sion," Christiansen said. "For example, if a designer knowingly places a gas tank on an automobile in a position that might result in a fatality in a minor accident, the public would be justified in considering that decision to be ethically questionable."

More subtle and challenging, however, is the fact that *no* engineering design is perfect. Unlike with mathematical equations, "there is no single correct solution to a design problem," said Christiansen. Moreover, "failures may occur even when we believe we have taken all possible pains to isolate or eliminate known failure mechanisms, confirmed adequate safety factors, and clearly defined the acceptable limits of customer use."

This difficulty becomes even more intractable with increasing complexity. "Ethical dilemmas lurk in the design, manufacture, and operation of complex systems," Christiansen asserted. "As systems become more complex and powerful, they also become less forgiving of mistakes. We try to postulate all possible failure modes in a piece of equipment, but for very complex systems, that is impossible. Failures can result from unpredicted interactions between complex parts of a complex system. Even so, when such a system fails, or when an equipment failure results from an honest engineering misjudgment, others may construe it as an ethical lapse.

"I am often asked about the value of engineering codes of ethics. Many engineers employed by corporations—with the possible exception of civil engineers working directly on public projects—are unaware of the content or even the existence of professional codes of ethics," he said. While he does not dismiss the purpose of formal codes, in his opinion "they are useful primarily as guidelines." Citing the first article in the IEEE Code ("We . . . agree to accept responsibility in making engineering decisions consistent with the safety, health, and welfare of the public, and to disclose promptly factors that might endanger the public or the environment"), "the devil is in the details," he warned. "Disclose to whom? Also, disclosure does not guarantee action or solution of the problem." So, in his view, each ethical situation must be dealt with on its own merits.

cially approves of undergraduate-level courses that focus on ethics in engineering and engineering management.

"Before Three Mile Island, the Challenger tragedy, and Chernobyl, professional ethics and ordinary morality were often deemed to be equivalent, and the province of philosophers and theologians," he recounted. In his view, students need to be steeped in "an understanding of when and how ethical issues arise in the everyday routine of engineers," and to "explore the impacts of corporate culture and peer influences on ethical behavior." [For his detailed thoughts, see his box "Engineering ethics—who cares?" above].

Learning and teaching the many nuances of ethics in engineering is a difficult challenge, Christiansen admitted. It may even be unrealistic to expect corporations to become active players in the process: "The very word 'ethics' can set the hearts of corporate lawyers racing as they envision expensive lawsuits and bad publicity. 'What our customers and the public don't know can't hurt us' may still play well within many companies, at least in the short run.

"But some CEOs are seeing that course as an archaic and dangerous path to follow. A few corporations are bringing ethics out of the closet, discussing specific ethical issues with their employees on a regular basis," he said. Still, the brunt of the educational task will probably fall to engineering schools, at least for the foreseeable future. "Nearly all the top research universities now offer courses having a significant engineering ethics content. But only 25 percent of the engineering undergraduates of all engineering schools, taken as a whole, are required to take such a course. I would suggest that, as we enter the next millennium, this percentage must and will increase dramatically.

"Of course, design is not the only phase of engineering in which ethics can rear its head," Christiansen said. "But I believe that if engineers, engineering managers, faculty, and corporate management focus on the potential ethical dimensions of design decisions, we will have taken an enormous step forward."

◆

Donald Christiansen, who received his Fellow award in 1981 "for contributions to professional communication in electrical and electronics technology," writes frequently on engineering topics, including safety, reliability, and ethical issues. He is the editor of Engineering Excellence: Cultural and Organizational Factors *(Piscataway, N.J.: IEEE Press, 1987).*

Christiansen also sees a potential source of wisdom for the future from an unexpected quarter: the study of engineering past. "Historians can hold up a mirror so that we see ourselves and our profession more objectively," Christiansen said. "Over the years, scholarship in the history of technology has turned from the straightforward documentation of technical developments (when, where, what, and who) to the inclusion of analysis of the influences of culture, the environment, economics, politics, and personalities on a particular technical development.

"As historians reveal more and more of how technology has been developed and exploited, they become the custodians of a body of knowledge that could be of great value to today's engineers," Christiansen pointed out. "Among other things, this approach opens up additional paths to the study of technological innovation and how better to innovate. Aside from providing examples of successful design approaches, an awareness of the history of technology can help avoid reinvention. Also, engineers can learn from the failures of others."

New flexibility in accreditation requirements may not, however, offer solutions to some contentious problems. For example, the compartmentalization of engineering design and device physics concerns Christiansen. "Engineers in the next decade may think of design too abstractly," he worried, saying he would like to see electrical engineers schooled in the physics of electron and electro-optic devices, as well as in the fundamentals of reliability engineering.

In sum, Christiansen said, dynamic developments in the twenty-first century mean that "electrical engineering will probably differ so greatly from what it has been historically that it will attract a different breed of student."

Catching Engineers Young

All the interviewees noted that if future generations are to become interested in science and technology, they need to experience the excitement of it long before they have reached university age.

To that end, Donna Shirley, the JPL's former manager of Mars exploration, is the spokesperson for the Mars Millennium project, a major education project for students from kindergarten through high school (*http://www.mars2030.net*). An official White House Millennium Project, it is jointly sponsored by the National Endowment for the Arts, the J. Paul Getty Museum in Los Angeles, the U.S. Department of Education, and NASA/JPL. "This is the first time so many government agencies have been involved in a project like this," she said. Moreover, companies such as America Online and the candy company Mars Inc., and organizations such as The Planetary Society, are helping to provide information, produce the Web site, posters, and other materials, and to celebrate the students' Mars community designs.

"The objective of the Mars 2030 project is to get groups of children and their teachers—or the Girl Scouts, the Boy Scouts, YMCAs, or whatever—to design a village for 100 people on Mars in the year 2030," Shirley explained. "We picked 2030 because kids who are 10 to 18 years old could possibly be in a community on Mars—they may actually be designing their future workplace and home."

The Mars Millennium Project differs in one key way from previous projects in which NASA has encouraged children to design a Mars habitat. The earlier projects have been "almost exclusively technical," Shirley noted, in which students were

Jeff Campson, marine life curator at the National Maritime Center in Norfolk, Virginia, builds the first underwater LEGO® city on World Oceans Day, June 7, 1997, to launch the LEGO® Deep Sea Challenge national building contest. The program, for children ages 5 to 12, is designed to stimulate their interest in ocean exploration.
Feature Photo Service

encouraged to consider how to get the colonists to Mars, protect them against the cold and the lack of oxygen, and grow enough food for them to survive.

The Mars Millennium Project, however, is multidisciplinary. The project asks students to consider "not only how to survive physically but also how to survive intellectually—to investigate all aspects of an actual, long-term community," she explained. In addition to the Web site and printed materials, part of the project

The Internet-based Mars Millennium Project (*http://www.mars2030.net*) will engage students from across the world in designing a Mars habitat that they might grow up to build. The Web page branches into various design areas, including social sciences, and resources. *Mars Millennium Project*

information will be a set of videotapes, moderated by Shirley, that feature dialogues between artists and scientists on the visual arts, architecture, dance, and music, to "bring together the creative artistic with the creative technical," she said. "The tapes will explore such questions as: What are the colors on Mars? How would objects look different from how they look on Earth if you were trying to paint a Martian landscape?" Shirley said. "How would dance work in three-eighths gravity? How would music sound in the thinner air?" [For some of Shirley's convictions about encouraging the imagination, see her box "Why is diversity essential to sustaining creativity?" p. 278.]

The aim is to get students to think about "the aspects of community," she said, as well as the scientific and technical implications. As they build a visual representation of the Mars community (which could range from a physical model or photographs to a Web simulation, depending on the age and skills of the students), the hope is that the project also makes them "think about their own communities on the earth," Shirley said. "We want them to link back to their own environment so it is not just esoteric 'out there' kind of stuff."

Excellence and Equality

Similarly concerned with both young people and the aspect of community is John B. Slaughter, president of Occidental College in Los Angeles, California. Slaughter "strayed" from a strictly technical career to become the president of a liberal arts college, he said, because he feels that people in science and engineer-

DONNA SHIRLEY: Why is diversity essential to sustaining creativity?

JET PROPULSION LABORATORY

"Managing creativity is about taking a diverse group of people with diverse skills and getting them to produce a brand new thing," affirmed Donna Shirley, former manager of the Mars Exploration Program at the Jet Propulsion Laboratory of the U.S. National Aeronautics and Space Administration (NASA) in Pasadena, California. Shirley was the original leader of the team that designed and built the endearing, six-wheeled, microwave-oven-sized rover named *Sojourner Truth*. After being successfully landed on Mars by the Pathfinder mission in 1997, the rover captured such international attention and affection that it inspired toy manufacturers to produce replicas for children.

The principal innovation in the creation of *Sojourner*, the first wheeled vehicle to roll over the surface of Mars to take pictures and make rock assays, was making the rover as *low* tech as possible. "It was made to work by devising everything to be as simple as

ing all too often "have blinders on" about wider social issues. [For more insight into Slaughter's motivation, see his box "How can people learn to get along better?" p. 284].

Slaughter is helping to redirect the future—one young student at a time. "Undergraduates are at such an impressionable stage in their lives," he explained. "They're leaving home for the first time, they're trying to decide what to pursue as a career, they're interacting for the first time with people of different backgrounds—they're in a crucible. It's critically important that they receive high-quality education. But it's equally important that they have the support structures

we could get it," said Shirley. "It was still very complicated, but it required relatively few major technology
breakthroughs. We just got a lot of clever people together
to figure out how to do it."

In her 25 years of managing space endeavors, Shirley found that encouraging creativity is not a regular
part of engineers' classroom education, but is typically
learned on the job. With *Sojourner*, a project under a
tight budget and seemingly insurmountable technical
problems, she emphasized with her team of engineers
how to think and do things innovatively. Now retired
from NASA's Jet Propulsion Laboratory, she teaches those
lessons to commercial companies through her speaking engagements and consulting company, Managing Creativity, La Cañada, California, as she works on her doctorate in human and organizational development at The Fielding Institute, a distance-learning school.

"Everybody primarily thinks of the creative process as just coming up with a new

> 'One of Donna's laws is that diverse teams are creative teams. And a corollary is: creative teams are harder to manage than homogeneous teams.'

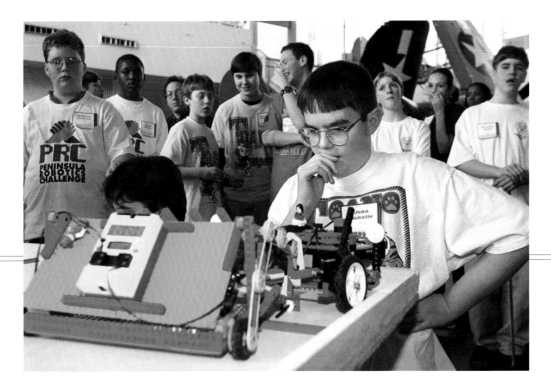

John Schulte, a student at Toano Middle School in Toano, Virginia, considers how to
get his robot out of a corner during the first annual Peninsula Robotics Challenge held
February 13, 1999, at the Virginia Air and Space Center in Hampton, Virginia. Students
in teams used LEGO® MindStorms, sophisticated toys with microprocessors, motors,
sensors, and other devices built in and connectable to each other. *NASA/Langley
Research Center*

'The drive to ever-shorter product cycle times doesn't allow scientists and engineers time for introspection, or for broadening themselves in a way that leads to different creative ideas.'

idea," Shirley said. "What I have is a process that focuses on creative processes through the product's full life cycle—on how you nurture the creativity of people as they take the initial idea, get alignment with what the customer wants versus what can actually be done, and produce a product—putting all the nuts and bolts together, marketing it, getting it out into the field, providing the warranty, and so on." Concentrating on life-cycle creativity "keeps people mindful of things that can kill the creativity at every turn."

Shirley's program has several parts. One is a set of lectures on the creative process and relating it to the Mars Exploration program. Another is a workshop in which she and a partner take a group of people and let them experience the process by working through creative solutions to a problem. "It is frightfully abbreviated, but we focus on getting people to understand why every individual is different and to appreciate the value of those differences to the creative process," she said. "We let them experience Donna's laws, one of which is that creative teams are diverse teams. And they also discover a corollary, which is that creative teams are harder to manage than a homogeneous team. If everybody on a team is just like you, it is comfortable. But if you have people who are different, your limits are stretched and the team is much more innovative."

that make it possible for them to become familiar with different cultures and ideas." And in an intimate college of only 1,600 students, he has found it is "quite possible to assist students individually in development and growth."

His mission at Occidental is to "prepare students for leadership, to recognize the importance of multiculturalism and inclusiveness, and to show that excellence and equity are not mutually exclusive." Although the college does not have its own engineering program, it does have a "three-two program with CalTech [the California Institute of Technology], in which students attend Occidental for three years and then transfer to CalTech to complete their engineering or physics de-

Diversity means not only being exposed to differing points of view of live humans, but also to differing intellectual approaches. In her experience, sustaining creativity means allowing time for daydreaming and for pursuing unrelated activities (such as art or music) that "open your subconscious up to connections," she said. "I read a lot eclectically, and I'm always finding connections that I normally wouldn't make if I were reading only in my field."

What troubles her, however, is that "with our current economy, which is focused on keeping stockholders happy, feeding the investment frenzy, and maintaining a 20-percent-per-year rate of return," the get-rich-quick mentality in commercial corporations "doesn't allow scientists and engineers time for introspection, or for broadening themselves in a way that leads to different creative ideas. The drive to ever-shorter product cycle times often results in burnout and errors as well as lack of creativity. Likewise, NASA's mantra of 'better, faster, cheaper' has produced some great missions, but it's being carried too far in some cases."

Over the long-term, she mused, "it will be interesting to see how it all plays out."

◆

Donna Shirley is the principal author of Managing Martians *(with Dannelle Morton; New York: Broadway Books, 1998), which recounts her own experiences of becoming a female engineer, working on a variety of space and energy projects, and finally—with* Sojourner—*overcoming technical and human challenges. She is currently negotiating with a Hollywood movie producer to turn the story into a television or theatrical motion picture. In 1997, she was named Woman of the Year by both* Glamour *and* Ms. *magazines. In 1999 she was selected to be assistant dean of engineering at the University of Oklahoma.*

gree," he said. And the college has the numbers as proof of performance: since Slaughter became its president in 1988, it has graduated three Rhodes scholars, three Marshall scholars, four Truman scholars, eight Goldwater scholars, 16 National Science Foundation Graduate Fellows, and 16 Watson Fellows.

But he also reaches down into high school and earlier grades. "We [at Occidental College] have a lot of interactions with public schools—most independent colleges do. I'm particularly involved in educational reform efforts to encourage young people of color to pursue science and engineering. Too many are discouraged from going into science and engineering because they don't have role mod-

Dr. Dannellia Gladden-Green of Texas Instruments Inc. is one of a new generation of engineers. The number of women and minorities in engineering, science, and technology is growing, bringing new perspectives and innovations to the industry. *PR Newsfoto and Texas Instruments*

els, haven't met engineers as they grew up, and still encounter counselors who do not encourage them to take the right math and science courses in high school," he said. "That happened to me in the 1950s," he added, wondering why the situation on the cusp of the twenty-first century is still so much the same.

"We need to rethink our long-term social priorities," Slaughter concluded. "Engineers need to get out of the comfort of their immediate environment and look out to see the needs of the broader world. They need to be *part* of the transformation of our society."

Predictions: University Revolution

"Engineering education [in a university] used to be a collection of scholars around a library," said Eleanor Baum, dean of the engineering school at the Cooper Union for the Advancement of Science and Art in New York City. But the combina-

Marc Benioff, senior vice president of Oracle Corp., joins fourth graders Marcus DeFrantz, left, and Ashley Green in installing computers on May 15, 1999, at Carver Elementary School in San Francisco. Carver is the 100th school to receive an entire network of computers as part of Oracle's Promise, the corporation's ongoing commitment to General Colin Powell's America's Promise program that is promoting literacy. *PR Newsfoto and Oracle*

tion of information and Internet technologies with "the outrage in society at the high cost of higher education" is going to "force traditional universities to rethink how they operate, and faculty how they teach," she declared. "I see *huge* changes coming in the next 10 years."

The changes will transform the classroom, the laboratory, the library, and the university at large—not to mention the way students are attracted [see her box "How can more young people be attracted to engineering?" p. 288]. "An engineering education used to consist primarily of the transmission of information from the faculty to the students," Baum explained. The ready availability of technical information on the Internet, however, reduces the importance of a central library. It also reduces the importance of a professor's role as lecturer. "There's more concern now with how to use information in design," she said, meaning that a professor now needs to be "much more of a mentor or facilitator."

Videoconferences and chat groups via the Internet allow students and professors to work together on projects—asking questions, discussing issues, and interacting at a distance, just as spontaneously and as readily as if they were sitting in the same classroom (a boon for, among others, handicapped or geographically isolated students). Technology also allows students to tailor their studies around their hours of employment, or to participate from different time zones. There are also mobile laboratories that allow far-flung students to conduct experiments, "and they're not that expensive," she added.

JOHN B. SLAUGHTER: How can people learn to get along better?

DON MILICI FOR OCCIDENTAL COLLEGE

"The key challenges for the twenty-first century are not technological ones," said John B. Slaughter, president of Occidental College in Los Angeles, California. "Clearly, there are issues facing us in healthcare and the environment. But I have little doubt they will be resolved. My main concerns for the next century center on our inability to get along with one another."

While Slaughter is concerned about human relations and human rights in a global context, "I'm more concerned about them here at home," he said. "The United States can't meaningfully assist in international human rights issues as long as it has domestic issues that are critical for it to resolve."

One of his primary concerns is how technology has "the potential of widening the gap between the haves and have-nots, which can make the income distribution difference in this nation even more of a major problem than it is today," he said. "Poor people don't have computers in their homes, but they still have children who go to public school. In the

Although the concept of distance learning is not new—after all, correspondence courses have been around for decades—the difference now is that it is "moving from the school of continuing education into the mainstream," Baum asserted. In her view, Internet technology means that there "is no longer a need to congregate in one place" for mainstream undergraduate and graduate studies. Moreover, although correspondence universities traditionally have not been heavy hitters in engineering, Baum sees that changing as the new educational players "address the quality of courses and accreditation issues."

Baum is not alone in her predictions. "Today's universities are a lot like a

schools, their children are competing with economically advantaged children who've grown up with computers and technology.

"Even worse," he continued, "in some poor districts not even the schools have computers for the children to learn and use. Thus, low-income children are technologically and educationally handicapped from birth through graduation, making it difficult for them to get jobs requiring high-tech skills. Thus, it is difficult for a person who is economically disadvantaged to escape his or her origins," Slaughter said.

"I'm further concerned that the opportunities for young men and women of color to go to college are being threatened by recent legislation in some states—already passed in Washington and California—to repeal or nullify the effects of affirmative action," he added, referring to the actions of businesses, educational institutions, and government agencies to take aggressive steps to overcome past racial and gender biases and underrepresentations of minorities and women in the workforce and higher education. "All that is going to do is make it even harder for young people of color to have even the

Familiarity breeds skill. With SketchBoard, a kids-level digitizing pad and software, children can learn to draw on the computer. A challenge facing society is ensuring that youngsters at all economic levels have access to such basic skill-building tools. *PR Newsfoto and Kid Board, Inc.*

potential, much less the means" to obtain an education necessary for the twenty-first century, he said.

"The situation becomes a positive feedback loop. And," he added gravely, "as anyone with an engineering background knows, positive feedback leads to instability—which is what we see in our society."

'When people have the opportunity to communicate ... the barriers disappear.'

In principle, the challenge is not insurmountable, as he has discovered over his 11 years at Occidental, a college that has distinguished itself for its ethnic, racial, and cultural diversity. Nearly half of the student body is composed of persons of color and half the faculty appointments since 1989 have also been persons of color. "One thing I've learned from being in a small school is that when people have the opportunity to communicate across would-be barriers of race, culture, and language, they establish understanding," Slaughter said. "And they discover that there is so much more they can do in collaboration. The level of communication is enhanced and the barriers disappear."

◆

John B. Slaughter, who received his IEEE Fellow award in 1978 "for contributions to the design of digital, sample-data control systems," is co-author (with Richard E. Lapchick) of The Rules of the Game: Ethics in College Sport *(New York: Macmillan Publishing Co., 1989).*

smokestack manufacturing plant of the 1960s," agreed Wade H. Shaw, Jr., professor of engineering and technology management at the Florida Institute of Technology in Melbourne, Florida. "They're bricks and mortar, they're in one location, they are hierarchical in organization, they *assume* what students need to know based on history and precedent, and they *assume* one size fits all. In addition, they are still filling 30 chairs all at the same time and taking four or five years to stamp students 'graduated' and send them out into the world."

But in the twenty-first century, that will have to change, he asserted. "Manufacturing industries have already figured out how to listen to what the customer

Students at the Florida Institute of Technology collaborate on a multimedia presentation in science education. Just as the military values sports for building future combat teams, engineers need to learn collaboration and teamwork early in their academic careers. *Florida Institute of Technology*

really wants, and how to make one of something at a time with special features. And the productivity gains have been tremendous. But the educational systems are not prepared for that. So, how can we convert our smokestack educational systems of the 1960s into more of a lean manufacturing philosophy?"

In his opinion, the rise in distance learning in mainstream universities is in response to another major need: the necessity of keeping experienced professionals current in their fields. "Practices in law and accounting don't change rapidly, but engineering practices do," he declared. "Keeping engineers current is a real problem.

"Historically, universities have not emphasized adult education and lifelong learning as part of their mainstream work," he continued. "But the demand now for career-related adult education is incredible. Companies are finding it is a lot cheaper to keep an initial employee current and competitive than to recruit a new one. But most traditional educational systems are not geared to handle that

ELEANOR BAUM: How can more young people be attracted to engineering?

"I think it is terribly important for the engineering profession to become more diverse, to encourage more women and minorities to think of engineering as a career," declared Eleanor Baum, dean of the school of engineering at the Cooper Union for the Advancement of Science and Art in New York City. "The reason is not just because it's morally the right thing to do. The real reason is that we can't afford to continue losing the creativity of such large groups of the population. Engineering is even starting to have problems in attracting the best and the brightest white males.

"The penalties for not reaching out to women and minorities will be very serious in the twenty-first century—not just for the engineering profession, but also for our economic well-being. One of the beauties of diversity is added creativity, new ways of looking at things. For example, many women look at product design very differently from many men. Incorporating their design considerations could make the difference between a product's succeeding in a market rather than missing a market," she pointed out.

"We've worked more than 20 years on attracting women to the profession, but the

demand." Hence the rise in alternative educational models and institutions, which are becoming more involved in life-long learning and specialized courses.

Shaw also feels it is increasingly important for colleges and universities to make a point of hiring faculty who have spent some years working in industry. "Engineering is the only profession taught by non-practitioners," he declared. "Can you imagine learning surgery in medical school by a surgeon who has never operated on a patient? Well, a significant fraction of engineering educators have not done real engineering. So professors may teach coding, for example, but never teach coding for *what*—for a real product."

numbers are still low. Women make up 16 to 17 percent of engineering graduating classes, but only 6 to 7 percent of working engineers. Minority registration stays embarrassingly low: although engineering enrollment is up for Hispanics, it is down for African Americans. Even worse, the interest of college freshmen in engineering has fallen off in the very best students. They're more interested in medicine, law, and business," Baum continued.

"I think the low numbers are in part tied to the public image of engineers, as revealed by popular television shows," especially shows that are popular with teenagers and young people still seeking direction for their lives, Baum said. "Lawyers have *L.A. Law*—look how glamorous the profession is portrayed to be. Physicians have *E.R.* [acronym for "emergency room"] and other shows where doctors and nurses are shown saving lives. But engineers have no equivalent. Thank goodness for *Star Trek*! *Star Trek Voyager* even has a female technology officer. But she's only half human—she's half Klingon—and *Star Trek* is only fantasy. It's not seen as being 'real' like *L.A. Law* and *E.R.*"

In Baum's view, the image of the engineer contrasts with the image of figures in Western society that teenagers consider heroes. "Young people don't see glamour and immediate social benefit in 'real' engineering. Engineers still have a nerd image, very much a working-class image where as employees they tend to be treated as commodities. Young people's heroes are in music and sports. The intellect is not as highly prized as other kinds of skills."

Engineers' image problem with young people means that "engineering colleges and

> 'Engineering colleges and societies need to reach out to pre-college students. Many young people are thinking about careers and professions by age 12.'

Finally, Shaw feels that universities and students "need to recognize that management skills are *part* of the technology. Learning management is as important as learning the core enabling technologies." [For Shaw's more detailed thoughts on this issue, see Chapter 3, Techniques and Systems.] "In fact, without management skills to cultivate, market, and transfer technology, all the resulting benefits risk fading away at great peril."

The rise of distance learning and for-profit educational institutions could pose a significant threat to traditional universities, Baum pointed out. In part, they open up the "bizarre possibility of the creation of educational superstars," she noted.

societies need to reach out to pre-college students," Baum said. "Traditionally children in K–12 [kindergarten through twelfth grade] have been mostly hands-off for engineering societies and colleges. But that's a mistake. Many young people are thinking about careers and professions by age 12. And a lot unwittingly cut off their options for college by not taking the right courses in high school—such as not taking math beyond plane geometry, because it isn't required for graduation. Engineers need to be willing to volunteer to speak to classes about their profession, and to help teachers prepare experiments—to be role models at all K–12 grades," she suggested.

"One problem is that engineers themselves are working the problem of their image. If you ask engineers what engineers do, often you get technobabble. What engineers need to say is, 'we do things to make life better for people—we're problem-solvers for the benefit of society.'

"Maybe they need outside help. For example, the Australian Institution of Accoun-

The importance of role models on television cannot be underestimated. In a tale recounted by Nichelle Nichols—who played communications officer Uhura on the original *Star Trek* TV series—in the 1960s, Caryn Johnson called to her mother, "Mama, there's black lady on TV and she's not a maid." Caryn grew up to be comedienne Whoopi Goldberg. Another young lady of color who watched the series was Mae Jemison. She grew up to become a medical doctor and an astronaut. Here she works with astronaut Dr. Jan Davis, an engineer, to prepare a life sciences experiment during the Spacelab J mission aboard the Space Shuttle in 1992. Jemison also played a bit part in an episode of *Star Trek: The Next Generation*. *NASA/Marshall Space Flight Center*

tants, concerned about the image of accountants as being boring pencil-pushers, contracted a public relations firm to do a series of television advertisements. One ad showed a limousine pulling up in front of a corporate headquarters and a well-dressed person— a man in some ads, a woman in others—walking down the hall, while a narrator told of how accountants can turn whole companies around.

"Now, that's dramatic stuff!" Baum exclaimed. "As nontraditional as it may seem, I think engineering societies should hire outside PR firms to promote the importance of engineering, with non-engineers writing the ad copy and doing the recruiting—because PR firms are in touch with what excites the minds of today's 15-year-olds.

"We're not going to solve the problem of how to attract more young people to engineering quickly. But we can start by upping the engineer's image. And we can make its human application clearer. For example, in those fields of engineering to which women are attracted in high numbers—environmental engineering and biomedical engineering— it is very clear how engineering is helping people.

"The connection of the ways in which engineers help society is too often missing when we talk about the profession. But that connection needs to be made."

◇

Eleanor Baum, who received her IEEE Fellow award in 1990 "for achievements and leadership in engineering education, and efforts to increase the number of women and minorities in the engineering profession," is past president of the American Society for Engineering Education (ASEE) and of the Accreditation Board for Engineering and Technology (ABET). She is chair of the New York Academy of Sciences and of the Engineering Workforce Commission of the Engineering Directorate of the National Science Foundation.

"What might happen, for example, if everyone learns thermodynamics from one particular charismatic professor on videotape?" she asked. Moreover, despite being for profit, new providers are often offering undergraduate and graduate education at a significantly lower cost than traditional institutions.

"I think all these changes will decrease the number of universities," Baum stated. "The marginal ones will fold. I think state universities will change as they're under pressure from state legislators to lower costs and improve efficiency.

"We're seeing a revolution here," she concluded. "The acceleration of this is difficult to conceive—but boy, oh boy, is it coming." ◇

APPENDIX: THE FIFTY TECHNOLOGY EXPERTS

Below, in alphabetical order, are brief biographies of the fifty technology experts featured in *Engineering Tomorrow: Today's Technology Experts Envision the Next Century*. Those designated with (F) are IEEE Fellows, an award of unusual professional distinction conferred only by the Board of Directors of the IEEE (Institute of Electrical and Electronics Engineers) upon a person of extraordinary qualifications and experience. Other member grades shown are member (M), senior member (SM), and Honorary Member (HM)—for a person elected by the IEEE Board of Directors from among those who have rendered meritorious service to mankind in engineering or allied fields.

Rod C. Alferness (F): Pioneer in integrated optic components and subsystems, as well as wavelength-division-multiplexed optical communications networks; chief technical officer of the Optical Networking Group at Lucent Technologies Inc. in Holmdel, New Jersey.

Ghassem Asrar (SM): Associate administrator for earth science at the National Aeronautics and Space Administration (NASA) headquarters in Washington, D.C.

Norman R. Augustine (F): Former CEO of Lockheed Martin Corp. (1987–95), recipient of the Presidential Medal of Technology, and lecturer with the rank of professor on the faculty of Princeton University's School of Engineering and Applied Science.

Eleanor Baum (F): First female dean of an engineering university; dean of engineering at the School of Engineering at the Cooper Union for the Advancement of Science and Art in New York City.

Robert A. Bell (F): Founder of the Research Department at the Consolidated Edison Company of New York, an activity he led for more than twenty-eight years, eventually rising to the level of vice president; currently retired in New York City.

Linda Sue Boehmer (SM): Champion of advanced technologies in the rail industry; owner, LSB Technology, Clairton, Pennsylvania.

Stewart Brand: Founder and editor of the original *Whole Earth Catalog* (1968–72) and of the *CoEvolution Quarterly* (1974–85), and author of numerous books; co-founder of consulting firm Global Business Network, Emeryville, California, and The Long Now Foundation, San Francisco, California.

A. Robert Calderbank (F): Co-inventor of space-time codes, breakthrough wireless technology that uses a small number of antennas to provide superior data rates and reliability, contributor to the V.34 high-speed modem standard, and a pioneer in quantum computing; vice president, Information Sciences Research at AT&T Laboratories, Florham Park, New Jersey.

S. Joseph Campanella (F): Inventor of echo cancellation and an early pioneer in the design of communications satellites; chief technical officer, WorldSpace Corp., Washington, D.C.

Vinton G. Cerf (F): Co-inventor, with Robert E. Kahn, of the Internet and of the TCP/IP protocol, for which both received the U.S. National Medal of Technology in 1997; senior vice president, Internet architecture and technology, MCI Worldcom Corp., Reston, Virginia.

Donald Christiansen (F): Editor and publisher (1971–1993) of *IEEE Spectrum* magazine, which captured four National Magazine Awards (the magazine equivalent of the Pulitzer Prize) under his leadership; president, Informatica, Huntington, New York.

Rui J. P. De Figueiredo (F): Pioneer in the application of spline functions (equations that describe the curves naturally assumed by a spline, an elastic metal strip formerly used by draftsmen to draw smooth curves through points) to pattern recognition, signal processing, and image compression; director of the Laboratory for Machine Intelligence and Neural and Soft Computing at the University of California at Irvine.

George L. Donahue: A lead architect in charge of designing and upgrading the infrastructure of the U.S. National Airspace System for all aviation, and former vice president of Rand Corp. (1989–94); Federal Aviation Administration visiting professor at the Institute of Public Policy and School of Information Technology and Engineering, George Mason University, Fairfax, Virginia.

Freeman J. Dyson: Pioneer quantum physicist, a pioneer in the search for extraterrestrial intelligence, and best-selling book author; professor emeritus of physics, Institute for Advanced Study, Princeton University, Princeton, New Jersey.

Sylvia A. Earle: Oceanographer and marine biologist, and holder of a world-record deepest solo ocean dive (1,000 meters in 1986) made by a human being; founder and director of Deep Ocean Exploration and Research, Oakland, California, and director of the National Geographic Society's 1998–2003 Sustainable Seas Expeditions.

Douglas C. Engelbart: Inventor of the computer mouse; president, Bootstrap Institute, Fremont, California.

Thelma Estrin (F): A pioneer in the use of digital computers for electroencephalography (electronic study of the function of the human brain), and co-builder in the early 1950s, in partnership with husband Gerald Estrin, of Israel's first computer; Professor-in-Residence Emerita, computer science department, University of California at Los Angeles.

Richard L. Garwin: Member of the U.S. President's Science Advisory Committee 1962–65 and 1969–72, and of the Defense Science Board 1966–69, and a co-author of many books; IBM Fellow Emeritus at IBM Corp.'s Thomas J. Watson Research Center in Yorktown Heights, New York.

Wilson Greatbatch (F): Inventor of the first successful implantable cardiac pacemaker, for which he was inducted into the National Inventors Hall of Fame in 1986 and was awarded the 1990 National Medal of Technology; retired in Akron, New York.

John G. Kassakian (F): Pioneer developer of power electronics and founding president of the IEEE Power Electronics Society (1988); professor of electrical engineering and director of the Laboratory for Electromagnetic and Electronic Systems at the Massachusetts Institute of Technology, Cambridge, Massachusetts.

Alan Kay: Inventor of the idea of overlapping windows as a graphical object-oriented interface for a computer and co-founder of Xerox PARC (Palo Alto Research Center); vice president of research and development for Walt Disney Imagineering in Glendale, California.

David A. Kay: Former United Nations Chief Nuclear Weapons Inspector, leading numerous inspections into Iraq after the Gulf War to determine Iraqi nuclear weapons production capability (and spent four days as a hostage of Saddam Hussein in a Baghdad parking lot); corporate vice president, Science Applications International Corp. (SAIC) and director of The Center for Counterterrorism and Technology at Science Applications in McLean, Virginia.

Myron Kayton (F): pioneer designer of avionic, navigation, communication, and computer-automation systems, including the Apollo Lunar Module guidance and control system; president, Kayton Engineering Corp., Santa Monica, California.

Samuel J. Keene (F): Involved in the development and refinement of helium-neon lasers for the first industrial market applications, such as supermarket scanners and high-speed computer printers; chief technical officer of Performance Technology Consultancy in Boulder, Colorado.

Jack S. Kilby (F): Inventor of the monolithic integrated circuit in 1958, for which he was inducted into the National Inventors Hall of Fame in 1982 and received both the 1969 U.S. National Medal of Science and the 1990 National Medal of Technology; retired in Dallas, Texas.

Bennett Z. Kobb (M): Co-founder of the Wireless Information Networks Forum, a wireless-communications industry trade group, author of *SpectrumGuide: Radio Frequency Allocations in the United States, 30 MHz–300 GHz,* and veteran observer of wireless communications and FCC rulings for technical publications; president, New Signals Press, Arlington, Virginia.

Ray Kurzweil: Inventor of a number of commercial firsts, including a print-to-speech reading machine for the blind, a text-to-speech synthesizer, and a computer music keyboard capable of accurately reproducing the sounds of the grand piano and other orchestral instruments; chairman and CEO of Kurzweil Technologies, Inc., Wellesley, Massachusetts.

Cato T. Laurencin: Pioneer in tissue engineering of musculoskeletal tissues, especially bone, and the transfer of such technologies from the bench to the bedside; Helen I. Moorehead professor of chemical engineering at Drexel University and director of the university's Center for Advanced Biomaterials and Tissue Engineering, Philadelphia, Pennsylvania.

Tingye Li (F): Pioneer contributor to laser mode theory; research leader and staunch advocate for amplified, wavelength-division-multiplexed systems for high-capacity, multi-channel, optical-fiber communication; retired in Boulder, Colorado.

Robert W. Lucky (F): Inventor of the adaptive equalizer, a key enabler of modern data modems; corporate vice president, Telcordia (formerly Bellcore Applied Research), Red Bank, New Jersey.

Wayne C. Luplow (F): Pioneer developer of digital high-definition television; executive director, digital business development and high-definition television, Zenith Electronics Corp., Glenview, Illinois.

Ralph C. Merkle (M): Co-inventor of public key cryptography and pioneer in molecular nanotechnology; research scientist, Xerox Palo Alto Research Center, California.

Gordon E. Moore (F): Co-founder (with Robert Noyce) of Intel Corp. in 1968, serving as its CEO 1975–87, and articulator of "Moore's Law" in 1965 that computer power doubles every year; Chairman Emeritus, Intel Corp., Santa Clara, California.

M. Granger Morgan (F): Pioneering educator and designer of university curricula in science, technology, and public policy; Lord Chair Professor and head of the Department of Engineering and Public Policy, Carnegie Mellon University, Pittsburgh, Pennsylvania, and co-director of the university's Center for Integrated Study of the Human Dimensions of Global Change.

George S. Moschytz (F): Pioneer in developing the theory and design of switched-capacitor filters, active filters, and filter synthesis techniques for communications networks; director of the Institute for Signal and Information Processing at the Swiss Federal Institute of Technology in Zurich, Switzerland.

Edward Alton Parrish (F): Pioneer in pattern recognition and image processing, and developer in the 1970s of a seminal real-time operating system for microprocessor-based systems, which was used worldwide in both academia and industry; president of Worcester Polytechnic Institute in Worcester, Massachusetts.

Arno A. Penzias (HM): Discoverer in 1965, along with Robert W. Wilson, of the 3 K microwave background radiation in the universe, the first observational evidence in favor of the Big Bang theory of the creation of the universe—for which Penzias and Wilson shared the 1978 Nobel Prize in Physics; senior venture partner, New Enterprise Associates, Menlo Park, California.

Roger D. Pollard (F): Best known for his work on microwave analysis and the characterization of microwave semiconductor devices; holds the Hewlett-Packard Chair in High Frequency Measurements, School of Electronic and Electrical Engineering, the University of Leeds, England, and is deputy director of the university's Institute of Microwaves and Photonics.

William F. Powers (F): Involved in the development of the Saturn booster guidance system, Apollo mission analyses in the 1960s, and the Space Shuttle in the 1970s; vice president of research, Ford Motor Co. in Dearborn, Michigan.

Elbert L. "Burt" Rutan: Pioneer designer of a series of innovative aircraft, including the *Voyager*, the first aircraft to circle Earth without refueling (December 1986); founder and president of Scaled Composites Inc., Mojave, California.

Donald R. Scifres (F): Demonstrated the first distributed-feedback injection semiconductor laser and in the first array of continuous-wave diode lasers; chairman of the board, president, and CEO of SDL Inc., San Jose, California.

Wade H. Shaw, Jr.: Pioneer in the application of Web technology for collaboration and asynchronous delivery of educational materials in the areas of engineering management, quality engineering, and simulation modeling; Chair of the Engineering Management Program at the Florida Institute of Technology, Melbourne, Florida.

Donna Shirley: Managed the Mars Exploration Program at the National Aeronautics and Space Administration's Jet Propulsion Laboratory in Pasadena, California, and was the original leader of the team that built Mars Pathfinder's *Sojourner* rover; president, Managing Creativity, La Cañada, California.

John B. Slaughter (F): First director of the National Science Foundation to hold a degree in engineering rather than in the pure sciences; president, Occidental College, Los Angeles, California.

Charles H. Townes (F): Inventor of the maser, co-inventor of the laser, and pioneer observer of microwave observations of interstellar gas clouds, leading to the discovery of the first molecules in space; co-recipient of the 1964 Nobel Prize in Physics; University Professor of Physics at the University of California at Berkeley.

Rao R. Tummala (F): Pettit Chair Professor and director of the Low-Cost Electronics Packaging Research Center at the Georgia Institute of Technology, Atlanta, and the first Endowed Chair Professor in Electronics Packaging at the University of Tokyo in 1999.

Joseph R. Vadus: Co-leader of the U.S.-France Cooperative Program in Oceanography, whose marine technology program led to discovery of the wreck of RMS Titanic; president, Global Ocean Inc., Potomac, Maryland.

Stephen B. Weinstein (F): Pioneer in discrete multitone (DMT) modulation and data echo cancellation techniques; Fellow in the C&C Research Laboratory of NEC USA, Inc., Princeton, New Jersey.

Victor Wouk: Pioneer in high-voltage DC equipment and instrumentation (1946–56), solid-state, off-the-line switching regulators (1956–72), and hybrid electric vehicles (since 1965); president, Victor Wouk Associates, New York, New York, and U.S. technical advisor to the International Electrotechnical Commission's committee on electric road vehicles.

Robert Zubrin: Best known as developer and advocate of the Mars Direct approach for sending humans to Mars; founder and president of Pioneer Astronautics, Inc., and co-founder and president of The Mars Society, both based in Indian Hills, Colorado.

ABOUT THE AUTHORS

JOURNALIST TRUDY E. BELL HAS WRITTEN about the physical sciences and technology for three decades. She has been an editor for *Scientific American* magazine (1971–78), a founding editor of *Omni* magazine (1978–79), and a senior editor for IEEE *Spectrum* magazine (1983–97). The author of seven books and some 300 articles—fifteen of which have won top journalism awards—she has also written for the National Aeronautics and Space Administration (NASA), the Congressional Office of Technology Assessment (OTA), the National Science Foundation (NSF), and the international management consulting firm McKinsey & Co. Both her A.B. (University of California, Santa Cruz, 1971) and her A.M. (New York University, 1978) are in the history of science. Ms. Bell lives in Lakewood (Cleveland), Ohio, with her daughter Roxana.

ERIC WAGMAN

JOURNALIST DAVE DOOLING HAS COVERED the space program and other technologies for thirty years. He has been science editor for *The Huntsville Times* (1977–85) and editor-in-chief of *Space World* (1981–84). He is a co-author, with Wernher von Braun and Frederick I. Ordway, of *Space Travel: A History* (1985). He is also a winner of the National Space Club's Press Award (1981) and Goddard Historical Essay Award (1983), the American Society of Mechanical Engineers' Ralph Coats Roe Medal for communications (1985), and a NASA Team Achievement Award for support on the Microgravity Science Lab 1 mission. Mr. Dooling is the principal writer for the People's Voice Webby award-winning *Science@NASA* website at NASA's Marshall Space Flight Center in Huntsville, Alabama, where he lives with his wife and daughter.

DAVE DIETER, *THE HUNTSVILLE TIMES*

ABOUT THE EDITOR

JANIE FOUKE IS PROFESSOR OF ELECTRICAL and Computer Engineering and Dean of the College of Engineering at Michigan State University at East Lansing. Prior to accepting this position in 1999, she was on the faculty of the School of Medicine and the Case School of Engineering at Case Western Reserve University, Cleveland, Ohio, for 18 years. Her research and consulting activities focus on instruments and techniques to evaluate the pulmonary system. A senior member or Fellow of several professional societies, she is the author of more than 100 publications, including peer-reviewed manuscripts, book chapters, and proprietary reports. Dr. Fouke has recently completed a detail to the National Science Foundation in Washington, D.C., where she served as the first Director of the Division of Bioengineering and Environmental Systems within NSF's Engineering Directorate.

ACKNOWLEDGMENTS

EVERY BOOK IS THE PRODUCT of many hands. This one represents the dedication of an extraordinarily large number (hundreds) of people.

Chronologically speaking, credit and thanks go to Richard D. Schwartz, staff executive for business administration at IEEE, who provided the original vision and impetus for Engineering Tomorrow. Jonathan Dahl, staff director for marketing, refined the book proposal and presented it to the volunteer Technical Activities Board. Thanks are also due to the Technical Activities Board itself, and the 40 Societies and Technical Councils. The project would also have not been possible without the support of Anthony Durniak, staff executive for publications, and Daniel J. Senese, executive director.

Credit for the orchestration of the myriad of day-to-day publishing intricacies is due to the staff of the IEEE Press in Piscataway, NJ. On the editorial side, particular recognition is due to Kenneth Moore, director; Karen Hawkins, executive editor; Robert Bedford, staff assistant; and Franklin Dickson, intern. On the marketing and publicity side, recognition is due to Judy Brady, project manager, marketing; Michael Petro, key accounts sales manager; and Barbara Soifer, Press marketing manager; plus due to freelance agents Dudley Kay, SciTech Publications Inc., Mendham, NJ, and to Frank Tooni, RosicaMulhern & Associates Inc., Paramus, NJ.

As a coffee-table picture book is an unusual bird compared with the traditional technical publications of the IEEE Press, thanks for the innovative layout and design of the various editions are due to Laing Communications Inc., Redmond, WA, and Edmonton, Alberta, Canada. Sandra Harner, art director, and Kelly C. Rush, staff artist, were responsible for the visual presentation, while Laura B. Fisher, editorial/production manager, handled production coordination and style editing.

For their support and helpfulness in introducing us to the technology experts willing to be interviewed, gratitude is due to all the presidents of the IEEE Societies and Technical Councils:

IEEE Societies & Technical Councils	President
IEEE Aerospace and Electronic Systems Society	Myron H. Greenbaum
IEEE Antennas and Propagation Society	Daniel H. Schaubert
IEEE Broadcast Technology Society	Garrison C. Cavell
IEEE Circuits and Systems Society	George S. Moschytz
IEEE Communications Society	Tom J. Plevyak
IEEE Components Packaging, and Manufacturing Technology Society	John W. Stafford
IEEE Computer Society	Leonard L. Tripp
IEEE Consumer Electronics Society	James O. Farmer
IEEE Control Systems Society	Stephen Yurkovich
IEEE Council on Superconductivity	Moises Levy
IEEE Dielectrics and Electrical Insulation Society	James E. Thompson
IEEE Education Society	Karan L. Watson
IEEE Electromagnetic Compatibility Society	Dan D. Hoolihan
IEEE Electron Devices Society	Bruce F. Griffing
IEEE Engineering Management Society	Cinda S. Voegtli
IEEE Engineering in Medicine and Biology Society	Banu Onaral

IEEE Geoscience & Remote Sensing Society	Nahid Khazenie
IEEE Industrial Electronics Society	James C. Hung
IEEE Industry Applications Society	Ira J. Pitel
IEEE Information Theory Society	Ezio Biglieri
IEEE Intelligent Transportation Systems Council	Umit Ozguner
IEEE Instrumentation and Measurement Society	Stanley R. Booker
IEEE Lasers & Electro-Optics Society	Hans Melchior
IEEE Magnetics Society	Edward Della Torre
IEEE Microwave Theory and Techniques Society	Edward A. Rezek
IEEE Nuclear and Plasma Sciences Society	Igor Alexeff
IEEE Neural Networks Council	Clifford Lau
IEEE Oceanic Engineering Society	Glen N. Williams
IEEE Power Electronics Society	Phil T. Krein
IEEE Power Engineering Society	B. Don Russell
IEEE Professional Communication Society	Roger Grice
IEEE Reliability Society	Kenneth P. LaSala
IEEE Robotics & Automation Society	Toshio Fukuda
IEEE Sensors Council	John Vig
IEEE Signal Processing Society	Leah H. Jamieson
IEEE Society on Social Implications of Technology	Gerald L. Engel
IEEE Solid-State Circuits Society	Lew M. Terman
IEEE Systems, Man, and Cybernetics Society	Richard Saeks
IEEE Ultrasonics, Ferroelectrics, and Frequency Control Society	John R. Vig
IEEE Vehicular Technology Society	A. Kent Johnson

For the book's core content, our supreme thanks go to the 50 luminaries prominently featured in the book—each of whom generously devoted at least a full day of their time in preparation for the interviews, the interviews themselves, correcting the draft manuscript, and providing photographs and supplemental materials. Special thanks are particularly due to Donald Christiansen, former editor and publisher of IEEE Spectrum, for being an informal advisor to the project, and for his many valuable tips and suggestions for content, form, and execution.

In addition to those 50 featured luminaries, there were even more than that number of equal stature who also contributed their insights—either in the form of additional interviews or in the form of reviews of the chapter manuscripts. Although space prevents listing each one's title, many of them are full professors or deans of engineering at their universities, or ranged from specialist to senior managers at their respective corporations. Those names, in alphabetical order, are: Metin Akay, Dartmouth College, Hanover, NH; John B. Anderson, Lund University, Lund, Sweden; Lawrence K. Anderson, Lucent Technologies, Albuquerque, NM; B. Michael Aucoin, Texas A&M University, College Station, TX; R. Jacob Baker, University of Idaho, Boise, ID; Roy Billinton, University of Saskatchewan, Saskatoon, Saskatchewan, Canada; Col. Billie M. Bobbitt (retired), Sidney, OH; Mathias Bollen, Chalmers University of Technology, Gothenburg, Sweden; Richard C. Booton, Boulder, CO; Joe E. Brewer, Joe E. Brewer, P.E., Palm Coast, FL; Anne Brinkley, ISM Center of Environmentally Conscious Products, Research Triangle Park, NC; Charles D. Brown, Martin Marietta Corporation (retired), now at Wren Software, Castle Rock. CO; E. Ryerson Case, E.R. Case & Associates, Agincourt, Ontario, Canada; James J. Coleman, University of Illinois, Urbana, IL; Kenneth Dawson, TRIUMF, Vancouver, British Columbia, Canada; Christian de Moustier, Scripps Institution of Oceanography, University of California, San Diego, CA; Harold L. Dodds, University of Tennessee, Knoxville, TN; Murray Eden, Massachusetts Institute of Technology (emeritus) Cambridge, MA; Ferial El-Hawary, BH Engineering Systems Ltd., Halifax, Nova Scotia, Canada; Mohamed El-Hawary, Technical University of Nova Scotia, Halifax, Nova Scotia, Canada; Lois R. Ember, Chemical & Engineering News, American Chemical Society, Washington, DC; David Fogel, Natural Selection, Inc., La Jolla, CA; Dudley Foster, Woods Hole Oceanographic Institute, Woods Hole, MA; Robert L. French, R&D French Associates, Nashville, TN; Gerald H. Gaynor, G.H. Gaynor Associates, Inc., Minneapolis, MN; Denis O. Gray, North Carolina State University, Raleigh, NC; Michael D. Griffin, Orbital Sciences Corporation, Dulles, VA; Sandeep K. Gupta, Colorado State University, Fort Collins, CO; David A. Hammer, Cornell University, Ithaca, NY;

George Hanover, Consumer Electronics Manufacturers Association, Electronic Industries Alliance, Arlington, VA; William J. Hazen, Best Power Technology, Inc., Framingham, MA; Brian Hentschel, Mid-Atlantic Bight National Undersea Research Center, Institute of Marine and Coastal Studies, Rutgers University, New Brunswick, NJ; Joseph R. Herkert, North Carolina State University, Raleigh, NC; Robert J. Herrick, Purdue University, West Lafayette, IN; Breck Hitz, The Laser and Electro-Optics Manufacturers Association, Pacifica, CA; Geza Joos, Concordia University, Montreal, Quebec, Canada; Stamatios Kartalopoulos, Lucent Technologies, Holmdel, NJ; Myron Kayton, Kayton Engineering Corp., Santa Monica, CA; Ron Lenk, Fairchild Semiconductor, Mountain View, CA; N.C. Luhman, Jr., University of California-Davis, Davis, CA; Peter D. Meyers, Princeton University, Princeton, NJ; Tom Mock, Consumer Electronics Manufacturers Association, Electronic Industries Alliance, Arlington, VA; Marie-Jose Montpetit, Teledesic Corporation, Kirkland, WA; Janice G. L.-G. People, Plainfield Public Library, Plainfield, NJ; Dev Raheja, Design for Competitiveness, Inc., Laurel, MD; R. Keith Raney, Applied Physics Laboratory, Johns Hopkins University, Baltimore, MD; Giorgio Rizzoni, DOE Gate Center of Excellence on Hybrid Drivetrains and Controls, Center for Automotive Research, Ohio State University, Columbus, OH; J. Reece Roth, University of Tennessee, Knoxville, TN; Manuel F. Rodriguez-Perazza, University of Puerto Rico, Mayaguez, P.R.; Edl Schamiloglu, University of New Mexico, Albuquerque, NM; Peter E. Smouse, Cook College, Rutgers University, New Brunswick, NJ; Cary R. Spitzer, AvioniCon, Williamsburg, VA; Pradip K. Srimani, Colorado State University, Fort Collins, CO; Mehmet Ulema, Daewoo Telecom, Ltd., Middletown, NJ; Cinda S. Voegtli, Emprend Inc., Los Altos, CA; William A. Wallace, Rensselaer Polytechnic Institute, Rensselaer, NY; David J. Wells, University of Houston, TX; George W. Zobrist, University of Missouri-Rolla, Rolla, MO.

For additional helpful assistance with reviews and/or feedback on the manuscript, particular thanks are due to The American Institute of Aeronautics and Astronautics (AIAA), Reston, VA; The American Society of Mechanical Engineers (ASME), New York, NY; and the Society of Automotive Engineers (SAE), Warrendale, PA.

On a personal note, principal writer Trudy E. Bell would like to thank the IEEE Press for the honor and opportunity of participating in such a stimulating and satisfying project. She also owes a deep debt of gratitude to her daughter, Roxana K. Bell, for her patient understanding of Mama's long work hours, and to her mother, Arabella J. Bell, for providing many hours of quality grandmother-granddaughter time for Mama to work without interruption.

Co-author and photo researcher Dave Dooling also thanks the IEEE Press; Joseph R. Vadus for help in acquiring oceanographic illustrations; the National Aeronautics and Space Administration, Department of Defense, Oak Ridge National Laboratory, for having the foresight to provide high-quality images via the Internet; the many sources credited in the book for supplying images. Special gratitude go to his wife Sharon for her love and invaluable help in transcribing interview tapes and fielding inquiries, and her incredible patience with his overtime schedule during writing and editing.

Editor Janie Fouke, although thankful to the people cited above, especially the talented professionals at IEEE, also thanks Trudy and Dave. Ever resourceful and always pleasant, they taught her a lot!

INDEX

A.D. Little, Inc., 194

Abner, Jeff, *11*

Accreditation Board for Engineering and Technology (ABET), 266, 268–69, 270, 271, 291

Advanced General Aviation Technology Experiment (AGATE), *187*, 188, 190

Advanced Research Projects Agency (ARPA), 3, 4–5, 24. *See also* ARPAnet; DARPA

Advanced Television Systems Committee (ATSC), 127, 129

Aerospace Corp., 156

AeroVironment Inc., 261

Africa, 138, 201, 228

agriculture, 52, 53, 220

Airbus Industrie Inc., *57, 183*

aircraft carriers, 253, 256, 257, 258

aircraft, xiii, *57*, 183, 193, 200, 296; building, *57*, 69; consumer, *187*; design, 69, 197, 263; flight recorders, 65, 171, 180; hijacking, 249; military, 245, 253, 257, *257. See also* airplanes

Air Force, U.S., 195, 258

airplanes, 175, 182–90, *187*, 198, 211, 246, 261

Alcor, 166, 168

Alferness, Rod C., 9, 106, *110*, 110–12, 113, 293

al-Qaddafi, Muammar, 262

Ambrose, Stephen, 95

America Online, 275

American Institute of Electrical Engineers, xiii

American Society of Engineering Education (ASEE), 268, 291

Ames Research Center, 50, 52, 155

Amoco Technology Center, *93*

Amundsen, Raold, 202, 204–5

Apple Computer Inc., 80, 85, *86*, 91, 141; Macintosh operating system, 86, 87

ARPAnet, 4–6, 84

artificial intelligence, 3, 27, 31, 91, 146, 166, 168, 180

artificial organs, 14, 145

Asrar, Ghassem, 216, *222*, 222–25, 293

AT&T Bell Laboratories, 17, 21, 86–87, 96, 105, 107, 118, 120, 293

AT&T Laboratories-Research, 105, 106, 107

AT&T, 112, 123

Augustine, Norman R., 241, 242, *244*, 244–46, 253, 257, 258–59, 293

Aun Shin Rikyo cult, 251

Australian Institution of Accountants, 290–91

Austria, 191

Autodesk, *83*

automobiles, xiii, 20, 58, 62, 134, *171*, 171–73, 174, *182*; airbags, 13, 177; emissions/pollution from, 23, 37, 46, 161, 172, 173; navigation, 135, 186; power electronics in, 31, 32, 34, 36, *38*, 171; safety, 176–77, *179*, 181, 273. *See also* cars; electric vehicles; hybrid vehicles

autonomous underwater vehicles (AUVs), 203, 206–8, *209*, 210–11; *Proteus*, 203, *211*

Babbage, Charles, 41

Baker, David, 199, 200

Barksdale, James, *87*

Bartz, Carol, *83*

Basov, Nicolai, 24

batteries, 32, 33, 52, 173, 174, 175, 226, 229; car, 34, 36, 171, 176; lead-acid, 173, 175; life, 10, 12, 13–14; lithium, 13, 173; nickel/metal hydride, 172, 173, 174, *174*; tritium, 10; zinc-mercury, 12

Baum, Eleanor, 266, 282–84, *288*, 288–91, 293

Baxter Novacor, 11

Bellcore Applied Research, 123, 126, 140, 295

Bell, Robert A., 35, 55, 56–62, *58*, 234, 293

Benioff, Marc, *283*

Berkeley Literature Library, 89

Big Bang theory, 3, 18, 19, 21, 296

biology, xiv, 23, 25, 48, 49, 145–69

biomedical devices, 3, 13

biomedical engineering, 9, 14, 145, 149, 291

Bloomberg Television, *139*

Boehmer, Linda Sue, 171, *178*, 178–81, 293

Bolt, Baranek, and Newman, 4, 5

Bootstrap Institute, 80, 82, 83, 84, 92, 294

brain, 72–76, 150–52, 153, 154, 156–58, *159*, 160, 165, 294

Brand, Stewart, xii, 216, 217–21, *218*, 293

BREED Technologies, Inc., *41*

Brittain, Teresa, *165*

Brown, Jeanette, *17*

Burke, Edmund, 3

Burroughs, Edgar Rice, 201

Bush, George, 96, 199, 200, 205

Calderbank, A. Robert, 94, 96–98, 101, 107, 118–19, *120*, 120–23, 293

California Institute of Technology, 101, 168, 173, 191, 280

Campanella, S. Joseph, 125, 135–38, *136*, 293

Campson, Jeff, *276*

Canada, 46, *202*

cancer, 1, 138, 157, 159, 163, 262

Carnegie Mellon University, 64, 132, 169, 180, 227, 234, 295

cars, 35, 49, 96, *135*, 175, 181, 184–86, 213; collision avoidance, 41–42; safety, 177. *See also* automobiles

Carver Elementary School, *283*

cellular phone, xiii, 80, 113, 114, 119, 120

The Center for Counterterrorism and Technology at Science Applications, 242, 254, 256, 294

Cepheid variables, 2

Cerf, Vinton G., 3, *4*, 4–10, 24, 26–28, 294

Chardack, William C., 12, 13

China, 5, 45, 46, 100, 112, 174, 175, 237, 253

Christiansen, Donald, 266, *272*, 272–75, 294

Chrysler, 172

Cisco Systems, 155

climate change, xi, 1, 59, 213, 217, 227, 228. *See also* global warming.

Clinton, William J., 4, 151

coal, 34, 52, 56, 58, 59, 62, 234

collective intelligence, 72, 138

collective IQ, 82–83

communications, xiii, xiv, 2, 56, 58, 73, 103–23, 139, 142, 293, 294; digital, 185–86; fiber optic, 44; group, 7; in product development, 64–66; interpersonal, xii, 12, 64, 286; mass, 161; multimedia, 77; networks, 3, 18, 56, 69, 140, 293, 295; on Mars, 26; use of lasers in, 25, 44; wireless, 69, 77, 103, 105, 107, 113–19, 130–31, 132, 293, 295; workplace, 267, 270, 272

computer chips, 3, 16, 33, 65, 79, 80–82, 84–86, 198

computer intelligence, 77, 158, 166–68, 178–80, 181

computers, xiii, 3, 31, 49, 56, 60, 77, 79–101, 126, 141, 151, 173, 177, 185, 207, 224, 266, *285*, 294, 295; access to, 284–85; cost of, 44; dependence on, 41, 57; imaging, 125; in defense/military, 244, 253; in medicine, 150; in transportation, 180, 186, 187; laptop, 80, 130, 173, 179; mainframe, xiii, 13; personal (PC), xiii, 6, 13, 18, 79–80, 81, 97, 99, 130, 131, 187, 220, 224, 284

computing, 104, 130; devices, 27; power, 36, 84–86, 92, 113, 215, 228

Comsat Laboratories, 125, 136

Consolidated Edison Company, 55, 56, 58, 62, 293

Cooper Union for the Advancement of Science and Art, 282, 288, 293

Corporation for National Research Initiatives, 4

cosmology, 3, 17, 19–21

creativity, 13, 63, 74–76, 77, 117, 162, 163, 167, 169, 277, 278–81, 288

cryonics, 165

culture, xii, 2, 46, 74, 188, 220, 221, 274, 280, 286; corporate, 273

CyberFlyer Technologies, *100*

DARPA, 24, 123, 237. *See also* Advanced Research Projects Agency

Davis, Jan, *290*

de Figueiredo, Rui J. P., 56, 72–77, *74*, 294

Deep Ocean Engineering Inc., 228

Deep Ocean Exploration and Research, 225, 228, 294

defense, 241, 244, 245, 246, 248, 260, 262. *See also* war/warfare

DeFrantz, Marcus, *283*

Dellinger, Sam, *119*

Delta, *165*

Deneb, *143*

Department of Defense (DoD), 3, 7, 24, 64, 238

design (product), xii, 64, 272–74, 275, 288

digital audio radio services: terrestrial (DARS), 125; satellite (SDARS), 125, 135, 138

Digital Domain, Inc., *125*

disabled, 133, 168. *See also* handicapped

diseases, xiii, 1, 10, 15, 146, 148, 151, 153, 157, 158

diversity, 74, 75, 77, 278–81, 288

DNA, 23, 143, 255; biochip, *157*

Donohue, George L., 171, 183–91, *188*, 294

Douglas Aircraft, 183

Drexel University, 145, 146, 147, 148, 149, 295

Duke Communication Services, *119*

Dyson, Freeman J., xi, 193, 194–98, *196*, 199, 294

Earle, Sylvia A., xi, xii, 203–5, 212–13, 217, 225–27, *228*, 228–31, 239, 294

Earth Observing System (EOS), 222, 224–25

EC 2000, 266, 269–71

economically disadvantaged, xi, 6, 90, 284–85

economics, 3, 26, 53, 141, 205, 274, 288; Internet issues, 4–6, 89; Nobel Prize in, 26; of energy generation, 36, 56, 208, 197, 232; of manufacturing, 48, 52; of space colonization, 50

economy, 42, 94, 220, 231, 248

education, xi, xiii, xiv, 44, 65, 90, 138, 220, 223, 271, 281, 287–91, 296; computer, 94–96; cost of, 268, 281, 283, 291; engineering, 70, 265–91; public, 201, 225; standards, 136–137; undergraduate, 278, 291

Edwards Air Force Base, *195*

Egypt, *1*

Einstein, Albert, 20, 75, 97, 191

electrical engineering, xiii, xiv, 31, 34, 60, 74, 76, 266, 275

electricity, 31, 58, 199; generation of, 36, 56, 229, 234, 236, 237, 239

electric motors, 32, 33, 143, 160, 176

Electric Power Research Institute, 62, 238

electric power, 52, 55, 60, 61, 62, 238; distribution of, xiii, 56

electric vehicles (EV), 36, 171, *172*, 172–76, *174*, 229, 296; power electronics in, 33. *See also* hybrid electric vehicles

electronic commerce, 4, 7–8, 139

electronic publishing, 10, 103, 111, 116

electronics, xiii, xiv, 22, 25, 31, 43, 55, 57, 165; components, 33, 69, 95; consumer, 130; cost of, 15, 16; disposal of, 128–30; in transportation, 171, 179, 183, 187, 188; manufacturing of, 16, 48, 99. *See also* power electronics

electronic trading, 7

energy, 33, 56, 172, 207, 237–38, 281; alternative sources, xi, 9, 35–37, 172, 174, 234, 237 (*see also* solar power; wind power; hydroelectric power; fusion; fuel cells); conservation, 228, 234, 237; demand for, 35; of universe, 19; pollution caused by, 130, 231; renewable, 238

Engelbart, Douglas C., 80, *82*, 82–84, 92–94, 294

entertainment, xiii, xiv, 125–43

environment, xi, xii, xiv, 1, 62, 160, 215–39, 273, 284; codes, xii, 268–69; education, 270; effects of

fossil fuels on, 35, 56, 172; effects of technology on, 128–30, 267; preservation, 98; technology's impact on, 100–101, 161, 171, 274
Environmental Protection Agency, 175, 225
erbium-doped fiber amplifier (EDFA), 110–12
Estrin, Gerald, 154, 294
Estrin, Thelma, 145, 146, 150, 153, *154*, 154–56, 294
Ethernet, 5, 6, 140
ethics, xi, xiv, 269, 272–74
extraterrestrial intelligence, 193, 198, 294
extraterrestrial life, 207
Exxon Corp., 218

Fassbinder, Carol, *271*
Federal Aviation Administration (FAA), 66, 189, 190, 294
Federal Communications Commission (FCC), 105, 114–17, 122, 125, 127, 130, 132, 295
fiber optics, 22, 25, 44, 107, 158
The Fielding Institute, 279
financial industry, xiii, 57, 61
Florida Institute of Technology, 55, 68, 70, 71, 267, 286, *287*, 296
Food and Drug Administration (FDA), 152
food, xiii, 51, 53, 210, 218, 276
Ford Motor Co., 171, 173, 177, 180, 184, 185, 186, 296
Foresight Institute, 32, 50
fossil fuels, 1, 34, 51, 172, 173, 175; effects on environment, 35, 56, 58–59, 62, 234, 236; energy content of, 35, 36; prices, 34; supplies of, 174, 208, 232, 239. *See also* coal, oil, natural gas
France, 46, 95, 137, *183*, 210, 263
Franklin, Sir John, 205
Fraunhofer Institute, 137
Frazier, Don, *31*
frequency-division multiplexing (FDM), 136–37
fuel cells, 36–37, 173, 174, 237
Fullerton, Larry, 119
fusion, 29, 36, 232–34, 236–39
Fusion Technology Institute, 238

Gamow, George, 18, 21
Garwin, Richard L., xii, 241, 243, 246, *248*, 248–51, 253–54, 256, 294
General Mills, 34
General Motors Corp., 176
genetic engineering, 1, 9
George Mason University, 171, 183, 188, 294
Georgia Institute of Technology, 79, 80, 81, 98, 101, 296
geothermal power, 172, 234
Germany, 1, 5, 56, *57*, 137, 191, 245
Gibson, William, 41
Gladden-Green, Dannellia, *282*
Global Business Network, xii, 217, 218, 219, 293
Global Ocean Inc., 210, 296
Global Positioning System (GPS), 21, 135, 157, 164, 257
Globalstar, *103*, *104*
global warming, xi, 55, 59, 172, 215, 217, 234, 239. *See also* climate change
Goldberg, Whoopi, 290
graphical user interface (GUI), 80, 88, 92

Greatbatch Ltd., 14
Greatbatch, Wilson, xi, 3, *8*, 8–9, 10–14, 150, 232, 234, 236–39, 294
Great Britain, 255
Green, Ashley, *283*
greenhouse effect, 230, 235–36
greenhouse gases, 59, 172, 208, 229, 234

Hagen-Smit, Arie, 173
handicapped, xi, 145, 153, 270, 283, 285. *See also* disabled
Harvard University, 74, 151
heart, 9, 10–14, *15*, 150, *153*, 158, 165
Heinlein, Robert A., 77
Heisenberg, Werner, 76, 97
Honda Corp., 176
Honeywell, 253
Hoyle, Sir Fred, 143
Hubble Space Telescope, 2
Hudson Institute, 161
Hundt, Reed, 117
Hussein, Saddam, 242, 243, 252, 262, 263
hybrid electric vehicle (HEV), 171, 176, 296
hybrid vehicles, 32, *171*, 172, 174, 175, *176*
hydroelectric power, 36, 56

IBM Corp., xii, 45, 47, 81, 97, 99, 101, 106, 120, 248, 294
IEEE Aerospace and Electronic Systems Society, 299
IEEE Antennas and Propagation Society, 299
IEEE Broadcast Technology Society, 299
IEEE Circuits and Systems Society, 299
IEEE Communications Society, xiv, 134, 299
IEEE Components Packaging, and Manufacturing Technology Society, 299
IEEE Computer Society, xiv, 64, 96, 299
IEEE Consumer Electronics Society, 299
IEEE Control Systems Society, 299
IEEE Council on Superconductivity, 299
IEEE Dielectrics and Electrical Insulation Society, 299
IEEE Education Society, xiv, 299
IEEE Electromagnetic Compatibility Society, 299
IEEE Electron Devices Society, 299
IEEE Engineering in Medicine and Biology Society, 156, 299
IEEE Engineering Management Society, xiv, 71, 299
IEEE Geoscience & Remote Sensing Society, 300
IEEE Industrial Electronics Society, 300
IEEE Industry Applications Society, 300
IEEE Information Theory Society, 300
IEEE Instrumentation and Measurement Society, 300
IEEE Intelligent Transportation Systems Council, 300
IEEE Lasers & Electro-Optics Society, 46, 300
IEEE Magnetics Society, 300
IEEE Microwave Theory and Techniques Society, 300
IEEE Neural Networks Council, 300
IEEE Nuclear and Plasma Sciences Society, 300
IEEE Oceanic Engineering Society, 300
IEEE Power Electronics Society, 37, 294, 300
IEEE Power Engineering Society, xiv, 300
IEEE Professional Communication Society, 300
IEEE Reliability Society, 64, 300
IEEE Robotics and Automation Society, 300
IEEE Sensors Council, 300

IEEE Signal Processing Society, 300
IEEE Society on Social Implications of Technology, xiv, 300
IEEE Solid-State Circuits Society, 300
IEEE Systems, Man, and Cybernetics Society, 300
IEEE Ultrasonics, Ferroelectrics, and Frequency Control Society, 300
IEEE Vehicular Technology Society, 180, 300
implants: cochlear, 153–54, 156, 168; heart pumps, 9, 14, *15* (*see also* left ventricle assist device); neural, 1, 145, 158, 168; retinal, 158. *See also* pacemaker
India, 100, 174, 237, 245, 247
Informatica, 266, 294
Institute for Advanced Study, xi, 194, 196, 294
Institute of Electrical and Electronics Engineers (IEEE), xiii, 64, 71, 84, 210, 266, 293, 294; code of ethics, xiv, 273; Fellows, 6, 13, 27, 37, 42, 46, 62, 66, 76, 107, 123, 130, 134, 137, 142, 156, 163, 238, 246, 263, 270, 286, 291; publications of, 117, 142, 266, 272
Institute of Radio Engineers, xiii, 27
integrated circuit, xiii, 3, 12, 13, 14–17, 44, 80, 81, 96, 295
Intel Corp., 16, 79, 84, 85, 94–95, *95*, 96, 141, 271, 295
Interface Message Processor (IMP), 4
International Congress of Mathematicians, 96
International Mobile Telecommunications 2000 (IMT-2000), 113–14, 116
International Telecommunication Union, 113, 114, 136
Internet, xi, 3, 4–7, 76, 99, 103, 138–39, 141, 142, 151, 224, 243, 250, 277; access, 6, 89, 100, 103, 106, 107, 125, 126, 129, 130, 131; content on, 88–91, 110, 138; creation of, 3–7, 88, 294; government control, xiv, 5; group wisdom, 126; growth of, 105–106; home/consumer use, 95, 100; impact of, 4–10, 12, 44, 46; in education, 283–84, 90, 265; interplanetary, 24, 26–28, *28*; service providers, 103, 112; social relationships and, 132–34; software, 87, 91, 122; structure of, 72–74, *73*; traffic, 113, 118
Iran, 245, 247, 252
Iraq, 242, 245, 247, 254, 256, 257, 262, 294
Italy, 1

J. Paul Getty Museum, 275
Japan, 56, 98–99, 118, 133, 178, 186, 208, 210, *213*, 216, 227, 254
Jemison, Mae, *290*
Jet Propulsion Laboratory (JPL), 24, 28, 199, 203, 266, 275, 278, 279, 296
Jewish Hospital, 11
Johnson, Caryn, 290
Johnson Controls, 135, 179
Joint European Torus (JET), 236, 238
Julian, Percy, 151

Kahn, Herman, 161
Kahn, Robert E., 4, 5, 294
Kansai International Airport, 208
Kantrowitz, Arthur, 194
Kassakian, John G., 31, 32–38, *34*, 234, 294
Kay, Alan, 80, 86–92, *88*, 92, 294

Kay, David A., 242, 243, *254*, 254–56, 258–60, 262–63, 294
Kayton Engineering Corp., 242, 263, 295
Kayton, Myron, 242, 245, 251, 253, *260*, 260–63, 295
Keene, Samuel J., xii, 55, 61–68, *64*, 295
Keithley Instruments, *38*, *69*
Kilby, Jack S., 3, *12*, 12–13, 14–17, 295
Kimble, Jeff, 191
Kleinrock, Leonard, 4
Kobb, Bennett Z., 104–5, *114*, 114–17, 119–20, 122, 132, 134–35, 295
Kosko, Bart, 168
Krasnow, Lineene, *81*
Kraut, Robert, 132
Kulcinski, Gerald, 238
Kurzweil, Ray, 77, 146, 152–53, 157–58, *166*, 166–69, 295
Kurzweil Technologies Inc., 146, 152, 166, 295

Laboratory for Electromagnetic and Electronic Systems, 31, 34, 294
Laboratory for Machine Intelligence and Neural and Soft Computing, 56, 74, 294
Langer, Robert S., 150
lasers, *25*, 44, 47, 107, *121*, *131*, 196, 237, 295; accelerators, 197–98; gas, 44; in communications, 28; in medicine, *145*, *167*; invention of, 3, 22, 24–25, 296; propulsion, 194–95; semiconductor, 32, 43–47, 108, 109, 296; solid-state, 44, 46; space-based, 258; uses for, 22–24, 24–25
Laurencin, Cato T., 145, 146–51, *148*, 295
Lawrence, E. O., 196
Lawrence Livermore National Laboratory, 237
Lederman, Leon, 90
left ventricle assist device (LVAD), *11*, 14
LEGO®, 47, *276*, *279*
Li, Tingye, 105, *106*, 106–7, 111–12, 295
Lightcraft, *193*, 194, 195–96
Linux, 87, 91
Lockheed Martin Co., 200, 204, 244, 293
Long, Theresa, *69*
Los Angeles Metropolitan Transit Authority, *177*
Louis, Bob, *38*
Low-Cost Electronic Packaging Research Center, 79, 81, 98, 296
Lowell, Percival, 201
LSB Technology, 171, 178, 180, 293
Lucent Technologies Inc., 17, 106, *108*, 110, 113, 293
Lucky, Robert W., 123, 126, 138–43, *140*, 295
Lucles, Paul, 147
lungs, 15, 161–62, 262
Luplow, Wayne C., 125, 127–30, *128*, 295

management, xii, 26, 55, 67, 68–72, 219–20, 253, 265, 267, 272, 296
Managing Creativity, 279, 296
manufacturing, 32, 41, 95, 99–101, 129, 143, 188, 273, 286–87; chip, 81, 82; costs, 50, 52; molecular, 32, 47–53, 158
marketing, 19, 71, 89, 90, 91
Mars, 52, 193, *199*, 199–203, 296; exploration, 266, 275, 278, 280; *Global Surveyer*, 202; Internet, 26–28; living on, 52, 193, 201, 206, 208, 275–77; missions to, 28, 199–201, 204–6
Mars Direct, 199, 201, *202*, 203, 204, 296

Mars Inc., 275
The Mars Society, xi, *202*, 205, 206, 296
Marshall Space Flight Center, 31, *43*, *109*
Martin Marietta Astronautics, 200
Massachusetts Institute of Technology (MIT), 26, 31, 32, 34, 74, 147, 150, 151, 168, 178, 181, 294
MCI WorldCom Corp., 4, 6, 294
medical imaging, xiii, 115, 145, 158
medicine, xi, xiii, xiv, *22*, 25, 41, 49, 62, 109, 145–69, 227, 289
Medtronic Inc., 13
Merkle, Ralph C., 32, 47–53, *50*, 160–66, 168–69, 295
Metcalfe's law, 140–42
Metcalfe, Robert, 5, 140
micromachines, *48*, 260
Microsoft, 85, 90, 91, 122, 141; Windows operating system, 86, 87, 141, 253
microvehicles, 260, *261*
Microvision, *131*
military, xiii, 242, 245, 253
Milosovic, Slobodan, 259
minorities, xi, 149, 151, 281, 282, 285, 288–89, 291
Minsky, Marvin, 168
Mitsubishi Electric, 178
The Monster Board, *68*
Monterey Bay Aquarium Research Institute, 230
Moon, 27, *29*, 52, 184, 206, 208, *233*, 235, 239
Moore, Gordon E., 5, 79, 84–86, *94*, 94–96, 295
Morgan, M. Granger, 217, 227–29, 231–32, *234*, 234–38, 295
Moschytz, George S., xi, 146, 150–52, 153, 156–57, *160*, 160–63, 295
Moscow Aviation Institute, 199
Myrabo, Leik, 194, 196

nanomachines, *48*, *51*
nanostructures, 163–164
nanotechnology, 41, 50, 52, 160; molecular, 47, 49–53, 158, 160, 164, 165, 169, 295
nanotools, 49
National Academy of Engineering, 46, 91, 96, 201, 250
National Aeronautics and Space Administration (NASA), 52, *72*, 76, 93, 147, 159, 184, 187, 190, 195, 199–200, 205, 216, 222, 225, 239, 255, 278, 281, 293, 296. *See also* Ames Research Center; Jet Propulsion Laboratory; Marshall Space Flight Center
National Air and Space Museum, 223
National Endowment for the Arts, 275
National Geographic Society, xi, 225, 231, 294
National Institute of Standards and Technology, 225
National Inventors Hall of Fame, 8, 12, 232–34, 294, 295
National Maritime Center, *276*
National Oceanic and Atmospheric Administration (NOAA), 193, 217
National Research Council, 268
National Science Foundation, 81, 123, 266, 268, 281, 291, 296
National Space Society, 53, 206
National Television System Committee (NTSC), *126*, 127
National Transportation Safety Board, 181
NATO, 262

natural gas, 34, 56, 59, 239
Navy, U.S., 84, 205, 241, 246, *247*, 253, 254, 256, 257
NEC USA Inc., 126, 130, 132, 296
Netscape Communications Corp., *87*, 91, 122
New Enterprise Associates, 21, 296
New Signals Press, 104, 114, 117
Newton, Sir Isaac, 88
Nichols, Nichelle, 290
Nicholson, Dan, *141*
Nile, *1*
Nobel Prize, xi, 17, 18, 22, 24, 26, 90, 296
Norfolk International Airport, *100*
North American Security Dealers and Quotations market (NASDAQ), 7
North Korea, 241, 245, 246, 247, 253
NTT DoCoMo, *108*
nuclear power, 37, 52, 56, 58–62, 160, 172, 203, 234
nuclear reactor, 203

Oak Ridge National Laboratory, *55*, 145, *157*
Occidental College, xii, 277, 280–81, 284, 286, 296
ocean, 193, 210, 216, *217*, 217–20, 223, 235, 236, 237, 296; as life-support system, 215, 220, 228–31; communities, 193; exploration, 201, 203–11, 276; floating cities on, 208–210; on Europa, *207*; tourism, 211–13; underwater observatories, 225–27
oil, 34, 52, 59, 208, 219, 229, 234, 239
optical: character recognition, 146, 169; communications, 44, 107–9, 293, 295; signals, 108–9, 110; technology, 39; wavelengths, 40, 235
optical fibers, 103, 106, 107–10, *109*, 110, 112–13, 116, 118, 119, *121*, *159*, 161, 295
optics, 22, 25, 43, 69, 112, 131, 293; micro-, 46, 47; non-linear (NLO), 31
optoelectronics, 44, 108
Oracle Corp., 283
Orwell, George, 181

pacemaker, xi, 3, 8, 10–14, 38, 232, 294
Pakistan, 245, 247
Parcells, Dallas, *127*
Parrish, Edward Alton, xii, 265, 266–72, *268*, 295
Pathfinder: aircraft, 215; lander, 202
PennzEnergy Co., *35*
Penzias, Arno A., 3, 17–22, *18*, 296
Performance Technology Consultancy, xii, 55, 61, 64, 295
Philips Carin, 182
Pioneer Astronautics, Inc., 199, 206, 296
planes, 61, 62. *See also* aircraft
The Planetary Society, 275
Pollard, Roger D., xi, 31, 38–43, *40*, 296
pollution, xiii, 51, 100, 130, 160, 213, 215, 239, 270; air, 21, 161, 173, 174, 229; from automobiles, 37, 46, 176; ozone, 232
Potts, Zoe, *95*
Powell, Colin, 283
power electronics, 31, 32–34, 36–38, 56, 171, 237, 294
Power Harmonics, 153
Powers, William F., 171, 173, 175–76, 177, 181–82, *184*, 184–86, 296
Preskill, John, 101
Princeton University, 151, 194, 199, 236, 241, 293, 294

printed-circuit board (PCB), 80, 99
privacy, xiii, 2, 120, 130, 181, 243
processor, *16*
product development, xii, 64–66, 68
Prokhorov, Alexander, 24
prostheses, 145, 153, 154

quantum computing, 94, 96–98, 101, 123
Quinta, *97*

race, xii, 151, 285–86; discrimination, 148, 149–50, 151
radar, 38, 115, 120, 183, 186, 217, 245; military uses, 27, *241*, 254; spaceborne, 1; vehicular, 31, 41–42
radiation, 47, 60–61; 3 K microwave background, 3, 18, 19, 296
radio, xiv, 7, 95, 103, 116, 143, 183, 261; broadcasting, 134–36; digital, 125–26, 132–36 (*see also* digital audio radio services); frequency, 39, 40, 114, 119, 134; packet networks, 5–6; spectrum, 115, 119, 130, 132, 134; transmissions, *115*; transmitter, 14; ultrawideband (UWB), 119–20, 122; waves, 196–97
Reagan, Ronald, 260
Red Hat, Inc., 91
Reliability Society, 64
religion, 5, 74, 221
Remote Environmental Monitoring UnitS (REMUS), *226*
remotely operated vehicles (ROVs), *35*, 206–8, 210–11, *211*, 230; *Ventana*, 230
Rensselaer Polytechnic Institute, 194, 195, 196
research, xi, 17, 121, 201; corporate, 136–37; defense, 262; industrial, 119, 120–23; investment in/support for, 26, 106–7, 237–38, 272; market, 185; medical, 146, 148, 149, 150, 155, 156, 158; need for, 22, 237; outposts, 27, 203; software, 64; space, 212; university, 136–37
robotics, xiii, 49, *72*, 76, 205, 253, 279
rockets, 50, 104, *195*, 198, 203, 206, 246; Ares, 203; propulsion of, 29, 193, 194, 203; Saturn V, 171, 184
Rogers, Jan, *43*
Rotary Rocket, Inc., *195*, 197–98
Roton, *195*, 197–98
Russia, 216, 243, 245, 251, 252
Rutan, Burt, 193, 197–99, *200*, 200–201, 211–12, 296
Rutgers University, 225

Samsung, *63*
Satellite CD Radio Inc., 134
satellites, 26, *28*, 103, 113, *118*, 125, 129, 157, 185, 193, 215, 219, 225, 293; *Afristar*, 138; broadcasting, 134, 135, 136–138; *Corona*, 243; earth observation, 216, 217, 221, 222–25; *Earth Remote Sensing-1*, 217; geostationary orbit (GEO), 116–117, 223, 246; *Ikonos*, 243; intelligence, 241, 244, 262; low earth orbit (LEO), 104, *105*, 117–18, 212; navigation, 164, 181, *182*, 185, 186; packet networks, 5–6; spy, 242–43, *243*; surveillance, 248, 263; weather, 222, 223, 243
Saudi Arabia, 5

Scaled Composites Inc., 197, 201, 296
Schawlow, Arthur L., 24, 25
Schulte, John, *279*
Science Applications International Co., 254, 294
Scifres, Donald R., 32, 43–47, *44*, 296
SDL Inc., 32, 43, 44, 296
semiconductor lasers. *See* lasers: semiconductor
Sharp Electronics, *127*
Shaw, Wade H., Jr., 55, 68–72, *70*, 286–89, 296
Shepard, Alan, 212
Shirley, Donna, 266, 275–77, *278*, 278–81, 296
Shor, Peter W., 96, 97
signal processing, 33, 56, 76, 294
single-level integrated module (SLIM), 81–82
Slaughter, John B., xii, 5, 266, 277–78, 280–82, *284*, 284–86, 296
Smithsonian Institution, 228
Snow, C.P., 74
Society of Women Engineers, 180
software, 4, 69, 79–80, 85–88, 93, 99, 166, 250; architecture, 131, 134; costs of, 5, 52; dependence on, 61–62; design/designers, 85, 88, 92, 218; development of, xiv, 8, 63–65; engineers, 71, 85; market for, 90–91; open-source, 88, 91; operating system, 86–87; reliability, 62–64
Software Engineering Institute, 64
Sojourner rover, 202, 266, 278–79, 281
solar power, 36, 52, 138, 215, 234, 237
Solectria Sunrise, 172
Solow, Robert, 26
Sony, *131*, *133*, *141*
Soviet Union, 45
space, 27, 109, 194, 227, 255; -based defense/weapons, 242, 243, 258–59; colonization, 50, 193; earth observation from, 222–23; exploration, 53, 195, 206; medical experiments, 147, 164; missions, 28, 200, 223; tourism, 211–13
spacecraft, xiii, 69, 194, 195, 198, 200, 211, 263; *Deep Space I*, 199; *Galileo*, 207
Space Imaging Inc., 243
Spacelab program, 77
Space Shuttle, 1, *67*, 117, 203, 204, 239, *290*, 296
spectrum, 122, 221; applications for, 114–16; capacity, 123; infrared, 109; millimeter, 38–41; radio, 115, 119, 120, 132, 134; regulation, 123; submillimeter, 38–41; television, 125, 127; ultraviolet, 109
Stanford University, *155*
stock exchanges, 7, 9
Strategic Defense Initiative, 194, 244, 260
Stubbe, Federico, *141*
submersible vehicles, 201, 205
Sun, 28, 138, 172, 235, 236
Sun Microsystems, 87
surgery, xiii, 23, 145, 146, 149, 154, 156, 158, *159*, 288
Sweden, 56
Swiss Federal Institute of Technology, xi, 146, 150, 160, 295
Switzerland, 191

Taiwan, 174
Telcordia, 123, 126, 138, 140, 295

telephone, 6, 7, 44, 103, 113, 114, 125, 161;
  companies, 103; network, 77, 106; video, 163;
  wireless, 38, 103, 113, 114, 116. *See also* cellular
  phone
teleportation, 191, *191*
television, 6, 7, 44, 125, 127, 129–30, *139*, 162,
  163, 260; cameras, 254; companies, 103;
  competition with, 103, 116; digital, 114, 125–30,
  295; high-definition (HDTV), 17, *126*, *127*, 127,
  129, 130, 141, 295; programming/content, 88–89,
  103, 111, 289
terrorism, 245, 249, 254, 255
Texaco, *35*
Texas Instruments, 13, 282
Thermo-Electron, 11
Thomas J. Watson Research Center, xii, 120, 248, 294
tidal power, 172, 234
Time Domain Corp., *39*, *115*, 119
time-division multiplexing (TDM), 137
tissue engineering, 146–49, 150, 158, 295
Toor, Dev, 89
Toroidal Fusion Test Reactor (TFTR), 236, 238
Torvald, Linus, 87
Townes, Charles H., 3, 22–23, *24*, 24–27, 296
Toyota Corp., 32, 176, 187
Toyota Prius, *32*, *171*, *176*
TransGuide advanced traffic management system, *22*
transistors, xiii, 12, 15, 16, 17, 43, 79, 85, 86
transmission control protocol/Internet protocol
  (TCP/IP), 4, 6, 7, 141, 294
transportation, xiii, xiv, 20, 55, 62, 160, 171–91,
  194; public, 20–21, 178, 181. *See also*
  automobiles, aircraft
travel, 10, 182, 184, 250; business, 100, 130, 161,
  185; safety in, 189; space, 165, 194
Trinity College of Music, 74
Tucker, Dennis, *109*
Tummala, Rao R., 79, 80–82, *98*, 98–101, 296

U.S. Office of Management of Budget, 28
United Kingdom, 56
United Nations (U.N.), 256, 260
United States, 174, 175, 186, 188, 191, 206, 216,
  253, 284; CO_2 produced by, 237; communications
  in, 112, 125, 126, 132, 135; discrimination in,
  148; education in, 71, 83, 136, 268; military/
  defense, 241, 243, 246, 248, 250, 252, 258; power
  industry, 56; research in, 26, 106–7, 123; waste
  disposal/environment, 98, 100
universe, 18, 19–21, 197, 200, 296; creation of, 3;
  expansion of, 2
University of California: at Berkeley, 24, 27, 84,
  296; at Irvine, 56, 72, 74, 76, 294; at Los Angeles
  (UCLA), 3, 4, 145, 153, 154, 156, 294
University of Erlangen, 137
University of Leeds, xi, 32, 38, 40, 42, 296
University of Rochester, 237
University of Southern California, 155
University of Tokyo, 98, 296
University of Wisconsin at Madison, 154, 238
UNIX, 86–87

Unmanned Combat Aerial Vehicles (UCAVs), *258*
USSR, 245, 250

Vadus, Joseph R., 193, 207–10, *208*, 212, 296
vehicles, 100, 135, 172, 173; accidents, 148;
  underwater, 203, 206, 212, *226* (*see also*
  autonomous underwater vehicles). *See also*
  automobiles; cars
Veterans Administration Hospital, 12
Victor Wouk Associates, 171, 172, 174, 296
video, 77, 142, 255, 261; camera, 230;
  conferencing, 9, 22, 114, 283; gaming, 116;
  goggles/headset, *131*, *141*; telephony, 142, 163
Viking, 27
Virginia Air and Space Center, *279*
virtual reality, 7, 83, *93*, *143*, *155*, 158, 193, 201
Vista, *159*
Vo-Dinh, Tuan, *157*
Voyager, 193, 200, 201, 296

Wachter, Eric, *145*
Walch, Stephen, 32, 50
Walt Disney Imagineering, 80, 86, 88, 294
war, xiv, 1, 241, 242, 243, 262
warfare, xii, 242, 245, 258, *259*, 260. *See also*
  defense
wavelength-division multiplexing (WDM), 107–8,
  111–13, 116, 293, 295
weapons, 241, 242, 244–53, 257, 262–63; biological,
  245, 248, 251; chemical, 245, *249*, 251, 262;
  neuro, 1; nuclear, 239, 241, 243, 245, 246, 248,
  250, 254, 256; space-based, 242, 258
Weinberg, Steve, 20
Weinstein, Stephen B., 126, 131, *132*, 132–34, 296
Weizmann Institute of Science, 155
Wilson, Robert W., 18, 296
wind power, 36, 234
wireless communications. *See* communications:
  wireless
Wireless Information Networks Forum, 105, 114, 295
women, xi, 150, 154–56, 282, 285, 288–89, 291
Worcester Polytechnic Institute, xii, 265, 266,
  268–70, 295
World Wide Web (WWW), 5, 6, 7, 17, *73*, 77, *100*,
  103, 110–12, 114, 126, 129, 130, 132, 139, 180,
  296; sites, 28, 275, 276, *277*
Worldgate Communications, *139*
WorldSpace Corp., 125, 135–38, 293
Wouk, Victor, 171, 172–75, *174*, 176, 296
Wright brothers, 195, 183
Wright, Orville, 211

Xerox Palo Alto Research Center (PARC), 5, 32, 47,
  50, 84, 91, 160, 294, 295
XM Satellite Radio, 134

Yazawa, Tak, *133*
Young, Henry, 147

Zenith Electronics Corp., 17, 125, 127, 128, 130, 295
Zubrin, Robert, xi, 193, 199–203, *204*, 204–6, 296